FACILITATING PERSONAL GROWTH
GROWTH
IN SELF AND OTHERS

VONDA OLSON LONG
University of New Mexico

Brooks/Cole Publishing Company

I(T)P ™ An International Thomson Publishing Company

Pacific Grove • Albany • Bonn • Cincinnati • Detroit • London • Madrid
Melbourne • Mexico City • New York • Paris • San Francisco
Singapore • Tokyo • Toronto • Washington

Sponsoring Editor: *Lisa Gebo*
Advertising Project Manager: *Romy Fineroff*
Editorial Assistant: *Patsy Vienneau*
Production Editor: *Marjorie Z. Sanders*
Manuscript Editor: *Patterson Lamb*
Permissions Editor: *Cathleen S. Collins*

Interior Design: *CompuKing*
Cover Design: *Roy R. Neuhaus*
Art Editor: *Lisa Torri*
Cartoons: *Mark Taylor*
Typesetting: *CompuKing*
Printing and Binding: *Malloy Inc.*

For more information, contact:

BROOKS/COLE PUBLISHING COMPANY
511 Forest Lodge Road
Pacific Grove, CA 93950
USA

International Thomson Editores
Campos Eliseos 385, Piso 7
Col. Polanco
11560 México D. F. México

International Thomson Publishing Europe
Berkshire House 168-173
High Holborn
London WC1V 7AA
England

International Thomson Publishing GmbH
Königswinterer Strasse 418
53227 Bonn
Germany

Thomas Nelson Australia
102 Dodds Street
South Melbourne, 3205
Victoria, Australia

International Thomson Publishing Asia
221 Henderson Road
#05-10 Henderson Building
Singapore 0315

Nelson Canada
1120 Birchmount Road
Scarborough, Ontario
Canada M1K 5G4

International Thomson Publishing Japan
Hirakawacho Kyowa Building, 3F
2-2-1 Hirakawacho
Chiyoda-ku, Tokyo 102
Japan

Printed in the United States of America

10 9 8 7 6 5 4 3

Library of Congress Cataloging-in-Publication Data
Long, Vonda Olson, [date]
    Facilitating personal growth in self and others / Vonda Olson Long.
            p.   cm.
    Includes bibliographical references.
    ISBN 0-534-33870-4
    1. Self-actualization (Psychology) 2. Self-actualization (Psychology)—Problems, exercises, etc. 3. Self-perception.
4. Self-perception—Problems, exercises, etc. 5. Counseling.
I. Title.
 BF637.S4L66   1996
 158—dc20   95-23584
                                              CIP

Dedicated to fellow sojourners Mary, Ruth, Julie, Janet, Reyna, Arlena, Nelda, and Hilary, and all our fellow travelers on the journey of growth; with a special tribute to those with the desire and intention of helping fellow sojourners along the way—and to those who already recognize that they can help others travel only where they themselves have gone.

CONTENTS

◇ *Skill Development Section:*
A Framework for the Helping Process

♦ PART I

Purpose and Goals of Helping Relationships   3
   1   Helping Relationships   5
   2   Human Dimensions and Counseling Orientations   9
   3   Purpose and Goals of the Helping Relationship   11

♦ PART II

Outcome and Process Goals in the Helping Relationship   15
   4   Prerequisites for Effective Helping Relationships   17
   5   Client Outcome and Process Goals   19
   6   A Stage Structure for the Helping Process   23
   7   Helper Personal Growth   27

♦ PART III

A Stage Structure for the Helping Process:
Pre-Stage Attitudes of Rights, Respect, and Responsibility   31
   8   Beliefs Affect Attitudes   33
   9   Attitudes of Rights, Respect, and Responsibility   37
  10   Rights versus Control   41
  11   Respect versus Judgmentalness   45
  12   Responsibility versus Rescuing or Blaming   49

♦ PART IV

A Stage Structure for the Helping Process:
A Five-Step Communication Model   53
  13   Communicating to Facilitate Self-Understanding   55
  14   A Five-Step Communication Model   59

♦ PART V

A Stage Structure for the Helping Process: Stages and Skills   71
  15   Stage Structure Overview   73
  16   Stage 1 Helping Skills   75
  17   Stage 2 Helping Skills   91
  18   Stage 3 Helping Skills   117
  19   Integration and Application of Skills   123

Self-Development Section:
A Framework for Personal Growth

◆ PART   I

Personal Growth and a Positive Self-Concept: Goals and Dilemmas   131

   1   Personal Goals and the Pursuit of Happiness   133
   2   A Positive Self-Concept: Internal Judgment and Approval   139
   3   Conditioning toward External Judgment and Approval   143
   4   Resulting Issues: The Inherent Dilemma,
        Discrepant Self, and Lost Identity   159

◆ PART   II

Personal Growth Model: Developing a Positive Self-Concept   169

   5   Components of a Positive Self-Concept   171
   6   A Positive Self Concept: Self-Acceptance,
        Self-Esteem, and Self-Actualization   175
   7   A Positive Self-Concept: Developing Congruence,
        Competence, and Internal Control   183

◆ PART   III

Underlying Beliefs of a Positive Self-Concept:
Rights, Respect, and Responsibility   199

   8   Underlying Beliefs: The Foundation for a Positive Self-Concept   201
   9   The Right to Be Yourself   207
  10   Self-Respect   211
  11   Self-Responsibility   215

◆ PART   IV

Interpersonal Communication Skills   219

  12   The Complexity of Communication   221
  13   Components of Communication—Listening,
        Responding, and Expressing   227
  14   Mechanisms of Communication: Verbal
        and Nonverbal Behavior   237

◆ PART   V

A Stage Framework for Personal Growth   241

  15   A Stage Framework   243
  16   Stage 1: The Presenting Problem   247
  17   Stage 2: Underlying Issues   251
  18   Stage 3: Direction and Change   257
  19   Integration and Evaluation   269
      References   271

P R E F A C E

Facilitating Personal Growth in Self and Others is a manual intended to assist the facilitation of personal growth. *Personal growth* is defined here as the process of moving toward improved psychological health, as evidenced by self-fulfillment, life satisfaction, and the development of a positive self-concept. For consistency of terminology, *positive self-concept* is used to refer to the general outcome goal of personal growth in this manual.

The general goal of personal growth and the basic principles of growth are the same whether applied to the self or others. The general principles, goals, strategies, and exercises presented in this manual, therefore, are applicable to both the facilitation of personal growth in yourself (addressed here in the Self-Development section) and to the facilitation of personal growth in others (addressed here in the Skill Development section).

For readers completing this workbook, general objectives include the following:

1. Readers will be able to grow personally toward improved psychological health, as evidenced by the development of a more positive self-concept;
2. Readers will be able to develop communication and helping skills to increase their ability to facilitate personal growth in others.

Assumptions, Intentions, and Limitations

The intention of this workbook is to provide one possible framework for growth toward the development of a positive self-concept as well as one possible framework for the helping process of facilitating personal growth in others. Many other frameworks could also be used for the goals, growth, communication skills, or facilitative skills addressed here.

Exercises and skills in this workbook were developed for use with transitional, developmental, personal growth issues in nonpathological populations. They are applicable to the facilitation of both self-development and growth in others in a wide range of helping relationships.

The skills in the stage structure for the helping process are based on an eclectic structure. There is no presumption to reflect a particular theoretical orientation but rather to reflect a compatibility with a broad combination of theories. The stage structure is not meant to be a theory, but a framework for the helping process.

The terms *psychological health* and *positive self-concept* carry varying connotations culturally. They are used here with the express invitation to the reader to consider them in the broadest possible sense. They are defined, in general, as reflecting a self-assessment of satisfaction with one's being and life. Inherent in this definition is an implicit respect for individual values and diversity—whether measured in terms of culture, ethnicity, race, language, family, gender, spiritual orientation, ability/disability, socioeconomic status, education, sexual orientation, or any other means by which individuals define themselves. In addition, readers are invited to substitute terminology for outcome goals that offers a better philosophical fit for them personally.

Ethical standards, as outlined by the American Counseling Association (1988) and the American Psychological Association (1992), should be maintained at all times when using this framework in a professional helping relationship.

Structure

The workbook is divided into two major sections: Skill Development and Self-Development. Each section contains five parts. The skill development and self-development sections may be completed separately or in conjunction with each other.

Written in a workbook-style format, the manual includes content, examples, and exercises. Its purpose is to provide both an organized system for self-development through a practical guide for personal growth and a system of skill development using a framework for the facilitation of growth in others.

The manual can be used separately or in conjunction with the book *Communication Skills in Helping Relationships: A Framework for Facilitating Personal Growth* (Long, 1996). This book provides additional background information and expansion on the concepts and skills presented in the manual. Parts I through V of the manual generally correspond to Parts I through V of the book.

Acknowledgments

This workbook is the result of direct and indirect input from students, clients, friends, and colleagues. Appreciation goes specifically to Mark Taylor for his creative original cartoons, to Trish Stevens, Sheri Lesansee, and Keli Alark for their contribution to tables and figures, to Lynn McDonald for her research assistance and the development of exercise scenarios and to Larry Burlew, Barry University, for his expert review. I am especially grateful to Lisa Gebo and Marjorie Sanders for their shepherding of this project, and a special thanks goes to Gayle Griffith, without whose help this workbook would remain a bundle of yellow papers.

A Note to Workbook Readers and Instructors

This manual is a guide for the facilitation of personal growth. The section titled Self-Development is focused on promoting the reader's own growth, and the section titled Skill Development is focused on helping the reader learn to facilitate the growth of others.

In using this workbook, you will find that parts and chapters vary in length. This is consistent with the primary focus of the respective parts and allows for supplementary exercises and reading. Depending on your emphasis, you may choose to complete parts of the workbook over greater or lesser time periods. For example, you may spend only one week each, on Parts I and II, while spending eight to ten weeks on Part V.

Skill Development Section

The skill development section addresses purpose and goals of the helping relationship in Part I, presents a framework for facilitating growth in Part II, explores attitudes affecting helping skills in Part III, provides a communication model in Part IV, and looks at specific helping skills in Part V.

The skill development exercises may be used either as an integral part of class or seminar time, or as supplementary work. They may be completed during class, outside class, or during group laboratory practice periods.

The skill development section includes exercises that need to be completed in dyads, triads, or small groups. The exercises may be videotaped, if possible, to provide direct feedback.

Dyads. When exercise instructions direct readers to form a dyad, they should identify a partner with whom to have an interaction. One takes the role of helper; the other takes the role of client. The client shares an issue with the helper. After completing the exercise, the two reverse roles.

Triads. When exercises call for a triad, readers form groups of three. One takes the role of helper, one the role of client, and the third, the role of observer. The observer acts as a neutral, objective person who observes the interaction and takes notes relevant to the exercise. The three rotate roles and repeat the exercise three times so that each has the opportunity to be helper, client, and observer.

Issues. Some exercises ask readers to play the role of a client and share an "issue" with the teammate designated as their helper. Issues can be made up or real. Real issues obviously can range from the mild annoyance of trying to find a parking spot to major marital problems or domestic violence. The issues selected should be ones that readers feel comfortable sharing and that are appropriate to public disclosure in class and short time frames. The purpose of the dyads and triads is to provide practice in developing skills. The purpose of the exercises is *not* to provide personal counseling (although they may be helpful).

Giving Feedback. Exercises will direct readers to solicit feedback from each other following the completion of dyads and triads. It is important that feedback be given constructively. Constructive feedback is given in a way that does not judge the person, but rather invites that person to assess his or her behavior and the effect it has on others. Constructive feedback allows readers to review their goals, assess their behavior, and consider changing their behavior if it is not consistent with their goals.

The following are some criteria for useful feedback:

1. *It is descriptive rather than evaluative.* Having behavior described rather than judged allows individuals to hear the feedback nondefensively and to use it as they see fit. When evaluative language is avoided, the hearer has less need to feel self-protective. "I never got to make eye contact with you" is descriptive. "You were avoiding my gaze" is evaluative.

2. *It is specific rather than general.* To be told that one is "dominating" will probably not be as useful as to be told, "Just now when we were deciding the issue, I didn't feel I had a chance to speak." "Dominating" could also be interpreted as evaluative.

3. *It takes into account the needs of both the receiver and giver of feedback.* Feedback can be destructive when it serves only our own needs and fails to consider the needs of the person on the receiving end.

4. *It is directed toward behavior that the receiver can do something about.* Frustration is only increased when a person is reminded of some shortcoming over which he or she has no control.

5. *It is solicited rather than imposed.* Feedback is most useful when the receiver has formulated the kind of question that those observing can answer.

6. *It is well timed.* In general, feedback is most useful at the earliest opportunity after the given behavior (depending, of course, on the person's readiness to hear it).

7. *It is checked to ensure clear communication.* One way of doing this is to have the receiver rephrase the feedback after it is received to ensure that it corresponds to what the sender intended. When feedback is given in a training group, both giver and receiver have the opportunity to check with others in the group on the accuracy of the feedback, to decide whether this is one person's impression or an impression shared by others.

Self-Development Section

The self-development section addresses goals and dilemmas of personal growth in Part I, presents components of a positive self-concept in Part II, explores underlying beliefs affecting self- concept in Part III, considers interpersonal communication styles in Part IV, and presents a stage framework for developing a positive self-concept in Part V.

The self-development exercises are intended for readers to complete individually for their own information. With the exception of Part IV—Interpersonal Communication Skills, where exercises require readers to work together in dyads or small groups—exercises can all be completed individually.

While directions and/or selected exercises may be completed within a group context, the intention is for the content of these exercises to be the property of the reader. There is no expectation that readers would be evaluated or be required to share the specific content of the exercises either publicly or with the instructor.

Because the purpose of completing the exercises is to promote personal growth, readers should be prepared to do some personal introspection and to experience a certain amount of discomfort that inevitably comes with growth and change. If completing specific exercises triggers major distress, however, readers should be encouraged to consider terminating the exercise, and also to consider seeking professional assistance outside the class or seminar. Exercises are *not* intended to be used for personal therapy within a class or seminar setting.

VONDA OLSON LONG

*S*kill Development Section

A Framework for the
Helping Process

"Cheshire cat," she began, rather timidly, "would you tell me, please, which way I ought to go from here?" "That depends a good deal on where you want to go to," said the cat. "I don't much care where …" said Alice. "Then it doesn't much matter which way you go," said the cat.

(LEWIS CARROLL, *ALICE IN WONDERLAND*)

Purpose and Goals
of Helping Relationships

If you can't identify your purpose, and don't know what
you're trying to accomplish, how will you know whether,
and when, you accomplish it? Effective communication skills
are built upon a knowledge of what you're trying to do,
why you're trying to do it, and how you plan to accomplish it.

Helping Relationships

A mother of a teen-age daughter once came into my office in distress. "I'm trying to understand her," she exclaimed, "but she just won't listen to me!" "What have you done so far," I asked, "to try to understand her?" With exasperation the mother said, "I've told her over and over again that she'll never get anywhere if she doesn't study. Education is everything." "What have you done," I repeated, "to try to understand her?" "I've told her I'd help her with her homework. I've told her she'll be grounded if she skips class. "But what have you done," I gently persisted, "to try to understand her?" "Oh," she stopped, "I don't know what to do to understand her. I guess I really want her to understand me."

We communicate for one of three reasons: to have our perspective understood, to understand others' perspective, or to help others better understand their own perspectives. Communicating to help others gain a clearer understanding of their own perspectives is often stated as the purpose of communicating and is probably the most effective type of communication in helping relationships. Attempting to have our own position understood, however, is probably the most prevalent goal of communicating in actual practice.

❖ **EXERCISE 1:   COMMUNICATION GOALS**
Think about the most recent conversations you've had. What was your purpose in communicating? What proportion of the time were you focused on (1) expressing your perspective? (2) trying to understand someone else's perspective? (3) trying to help someone else better understand his or her own perspective? Was your focus consistent with your purpose?

Often when we say we're trying to understand or help someone, we really mean that we want that person to understand *us*. We're not clear about our goal. To determine whether we've accomplished a goal, we must first know what the goal is. To be effective in helping relationships, we need to clarify our goal.

HELPING RELATIONSHIPS

Helping relationships are those in which one person's role is to help others with their emotional, mental, physical, and/or spiritual well-being. Many relationships, such as parent-child, student-teacher, or counselor-client, could be classified as helping relationships.

❖ **EXERCISE 2: HELPING RELATIONSHIPS**

Identify relationships you are in that you would consider helping relationships. Are you the helper or the one being helped?

A professional helping relationship is one in which the professional helper has had training in helping skills and receives payment, either through fees or salary, for services. In this manual the term *counseling* is used when referring specifically to professional counseling relationships. The term *helping* is used more broadly to refer to helping relationships that include but are not limited to counseling relationships. By this distinction we show that although not all helping relationships are professional counseling relationships, all effective counseling relationships are helping relationships. Basic principles of counseling can be applied to all helping relationships.

COUNSELING

❖ EXERCISE 3:   COUNSELING DEFINED

Think about your definition of counseling. Write it below, addressing its process, goals, purpose, and parameters.

❖ EXERCISE 4:   HOW DOES COUNSELING DIFFER FROM CONVERSATION?

For our purposes, a suggested definition of counseling follows.

Counseling is (1) an intentional, systematic, and replicable, skill-based interactive process with (2) the goal of client growth toward improved psychological health, (3) the purpose of facilitating that goal, and (4) specific focal, ethical, and logistical parameters.

This definition provides a distinction between counseling and conversation. Specifically, counseling has specific focal parameters (the growth of the client), ethical parameters (adherence to the American Counseling or Psychological Association's ethical guidelines (American Counseling Association, 1988; American Psychological Association, 1992), and logistical parameters (specified meeting time and place, fee structure, referral system, and expectations and goals).

Human Dimensions and Counseling Orientations

The goal of the helping relationship is to facilitate personal growth toward improved psychological health. Growth and psychological health involve human development, accomplished through affecting, in some way, the basic human experience.

The basic human experience is made up of, but not limited to, three conscious domains of affective experience: behavior, cognition, and the domain of the unconscious. It follows, then, that major counseling theories would use these dimensions as the focus of their intervention. One way to organize counseling theories, which is used here, is to divide them into four major orientations based on their primary intervention emphases. The major orientations and their emphases are shown below:

Orientation	Emphasis
Humanistic	Affective Experience (feelings, emotions, existential *being*)
Behavioral	Behavior (actions, *doing*)
Cognitive	Cognition (thought processes, beliefs, *choosing*)
Psychodynamic	Unconscious

Individual clients differ, and the human dimensions most prominent in any given situation will vary. Consequently, the effective helper may appropriately draw from a broad combination of counseling and developmental theories. The eclectic framework for the helping process presented in this manual is compatible with the major counseling orientations. Before you use the manual, however, an important step for you is to clarify your personal philosophy in a way that makes it meaningful to you and usable by you in conjunction with this framework.

❖ **EXERCISE 5:   PERSONAL PHILOSOPHY OF HELPING**

Although you may not have your personal philosophy of helping completely clarified, you are probably in the process of developing it. For the purpose of establishing a baseline comparison, identify and briefly describe your personal philosophy of helping to whatever extent it currently exists. How do the four major counseling orientations relate to your current perspective on helping?

Purpose and Goals of the Helping Relationship

To know whether we've accomplished a goal, we must first identify the goal that we want to achieve. Many people struggle in their efforts to help others—whether these are their children, friends, family, partners, or clients. They say they don't know what to do to help. More accurately, they are probably not clear about their goal in helping. Without clarity of focus, how can they possibly know what to do to help?

Goals include both outcome goals and process goals. *Outcome goals* are what we want to accomplish—the end result, where we're going, what we're striving for. *Process goals* are the activities involved in accomplishing the outcome goals—the means, how to get there. Our *purpose* is why we're doing what we're doing. Unless we understand our purpose and our goals, the means of accomplishing them will remain elusive and unclear.

❖ **EXERCISE 6: PURPOSE AND GOALS OF HELPING**

What's the outcome goal of helping (what do you hope to accomplish)?

What's the process goal of helping (how do you hope to accomplish it)?

What's your purpose in helping (why are you doing what you're doing)?

GENERAL AND SPECIFIC GOALS

The general goal of the helping relationship, from the helper's perspective, is the growth of the client. General goals have been identified with such terms as *psychological health, self-realization, self-understanding, effective functioning, optimal development, self-actualization,* and a *positive self-concept.* General goals cut across a diversity of clients and client issues and are recognized prior to our meeting with the client.

❖ **EXERCISE 7: GENERAL OUTCOME GOAL OF HELPING**

Identify your general outcome goal in helping. Choose terminology such as psychological health, *or* effective functioning, *that fits with your philosophy.*

Specific goals are those that the client, or client and helper together, identify that are client-, situation-, and time-specific. Specific goals, such as becoming more assertive, more spontaneous, or less controlling, fit under the umbrella of the general goal. They cannot be identified prior to meeting with the individual client.

OUTCOME AND PROCESS GOALS

Outcome goals are what we are striving for. As helpers we identify general outcome goals, such as psychological health; together with our clients, we identify specific outcome goals, such as becoming more assertive. Specific outcome goals help accomplish general outcome goals.

Process goals are a means, an action, or an activity. Process goals may be purely a means—driving for the pleasure of taking a drive—or a means to an end—driving for the purpose of getting to a destination. To attain outcome goals, however, we must have process goals. The universal process goal in helping relationships is growth and change.

PURPOSE OF HELPING

The purpose of helping—why we're doing what we're doing—is to facilitate the identified goals. The identified goals, used here for the purpose of establishing consistent terminology, are growth (process goal) toward increased psychological health as evidenced by a positive self-concept (outcome goal).

Mary came in complaining of overeating. She explained that she gets depressed and eats to feel better. Instead, she ends up feeling worse. She wants help in changing her eating patterns and in feeling better about herself.

The general outcome goal in this case could be identified as improved psychological health as evidenced by a more positive self-concept. Specific outcome goals might be identified as increased self-esteem and improved eating patterns. Process goals to facilitate growth and movement toward the outcome goals, in this case, might include such client activities as exploring self-identity, assessing underlying beliefs about herself, and/or developing a plan regarding her eating habits. The purpose is to help facilitate Mary's growth toward the identified goals.

Outcome and Process Goals in the Helping Relationship

Once you've identified your purpose and goals—facilitating growth toward a more positive self-concept—you need to know, first, what you mean by them. What is a positive self-concept? Second, you need to know how to go about accomplishing them. How do you *develop* a positive self-concept?

Prerequisites for Effective Helping Relationships

Counseling is a process that is intentional, systematic, replicable, and skill based. We plan for the process to happen and actively facilitate it.

Intentional means we employ particular interventions for a purpose. They are not used as a matter of chance, luck, or accident. *Systematic* and *replicable* refer to a fairly predictable sequence: a beginning, middle, and finish. The process is based on replicable skills that are built on principles of human growth and development. *Skills* are purposefully employed to facilitate client growth toward identifiable goals.

The effective helper needs an organized framework from which to operate. In this manual, we suggest that you need the following three prerequisites for effective helping:

1. Purpose and Goal Clarification: *What* are you trying to do?
2. Philosophy of Growth: *Why* are you trying to do it?
3. Stage Structure for the Helping Process: *How* are you going to do it?

❖ **EXERCISE 8:   PREREQUISITES FOR EFFECTIVE HELPING**

Consider how each of the three prerequisites fits with your current perspective on helping:

1. *What are you trying to do? (What do you currently identify as your purpose and goals?)*

2. *Why are you trying to do it? (What's your current philosophy of growth?)*

3. *How are you going to do it? (What systematic, replicable, and skill-based framework do you currently use?)*

CHAPTER 5

Client Outcome and Process Goals

In Part I you were asked to identify your *general outcome goal* in helping (for example, psychological health as evidenced by a positive self-concept), *process goal* (such as personal growth), and *purpose* (to facilitate achievement of the goals). Note that a wide range of terminology is used for outcome goals, reflecting the many theories of counseling and human development. Choose terminology that has a philosophical and theoretical fit for you.

For our purposes, and for consistency of terminology, the term *psychological health as evidenced by a positive self-concept* has been arbitrarily selected as the general outcome goal for the client in this framework. *Growth* is used as the client process goal, and the *facilitation of growth* is identified as the purpose of helping. If these terms have a philosophical fit for you, feel free to use them as you develop your own framework for the helping relationship.

❖ **EXERCISE 9:   OUTCOME AND PROCESS GOALS**

Identify below the general client outcome goal and process goal you've selected.

General Outcome Goal: for example, a positive self-concept

Process Goal: for example, growth

The purpose of counseling is to help facilitate the attainment of these goals. If we expect to be able to do that, we need to know what the words mean. What is a positive self-concept? What is growth?

❖ **Exercise 10:** **Outcome Goal Terminology**

Define and describe your selected outcome goal terminology.

❖ **Exercise 11:** **Process Goal Defined**

What is growth? Give as complete a definition as you can. How does your definition apply to your own personal growth process?

As the facilitation of growth is the purpose of counseling, we must have a clear perspective, not only of what growth is, but of how growth takes place. Otherwise, how can we be consistently effective in facilitating it?

❖ EXERCISE 12:   PHILOSOPHY OF GROWTH

Briefly describe your philosophy of growth. Identify core elements of growth that are predictable and common to humans—yourself and others—of diverse backgrounds.

Definitions of a positive self-concept (p. 185) and growth (p. 187) are provided in Part II of the self-development section of this manual.

CHAPTER 6

A Stage Structure for the Helping Process

One of my students, who was studying personality theory and theories of counseling, came into my office and sat down heavily. She drew in a deep breath and exclaimed, "I don't think I can do this! I'm getting all this information, but I don't have a clue what to actually do when I sit down across from a client! I need to have organized in my head what I'm trying to do and how I'm going to do it." I agreed with her.

A STAGE STRUCTURE FOR THE HELPING PROCESS

After we identify and define our goals and clarify our philosophy of growth, we must ask, "How do we facilitate growth toward the accomplishment of the goals?" To work effectively and consistently, we need an organized structure, or framework, from which to operate.

❖ EXERCISE 13:   CURRENT PERSONAL APPROACH TO THE HELPING PROCESS

Briefly describe how you now organize your approach to helping.

In this manual, a Stage Structure for the Helping Process is presented for your consideration as you develop your own philosophy and approach to the helping process. The stage structure is based on hierarchical stages and addresses client process, helper skills, and relationship goals in each stage. Helper skills incorporate a five-step communication model and facilitative helping skills that correlate with each step (see Table 6.1)

Stage Focus	Client Process	Helper Skills		Helping Relationship Goals
Pre-Stage Assumptions: Underlying Beliefs & Attitudinal Goals	Underlying Beliefs: Rights Respect Responsibility	Attitudinal Goals: Rights Respect Responsibility		Resulting Rapport: Rights Respect Responsibility
Stage 1: Presenting Problem	Telling the Story Initial Awareness • Thoughts • Feelings • Behavior Initial Problem Identification	*Communication Model* 1. Listen 2. Center 3. Empathize	*Facilitative Skills* Attending Genuineness Positive Regard Boundary Distinction Explicit Empathy Concreteness	Rapport and Trust 3 R's Stage 1 Client Process Stage 1 Helper Skills
Stage 2: Underlying Issues	Problem Differentiation • External Disappointments • Internal Issues Examining the Issues New Perspective and Insights	4. Focus	Empathic: Implicit Empathy Confrontation Expressive: Self-Disclosure Immediacy	Processing and Understanding 3 R's Stages 1 and 2 Client Process Stages 1 and 2 Helper Skills
Stage 3: Direction and Change	Direction Implementation Change	5. Give Directional Support	Directionality Implementational Support	Directionality and Change 3 R's Stages 1, 2 and 3 Client Process Stages 1, 2 and 3 Helper Skills

TABLE 6.1
A Stage Structure for the Helping Process

The Stages

The stage structure is in three parts: Stage 1, where the focus is on the presenting problem; Stage 2, where underlying issues are explored; and Stage 3, where direction and change take place. In addition, pre-stage assumptions, based on underlying beliefs and attitudes, are addressed.

The Client Process

The client process offers general criteria for client focus as he or she moves through the three stages from the presenting problem to direction and change. Client process is addressed in Part V of the self-development section.

Helper Skills

The segment on helper skills provides a framework of skills for the helper to draw on as he or she moves with the client through the three stages of growth. The helper skills segment includes underlying attitudes affecting helping skills, a five-step communication model embedded in the three-stage helping structure, and corresponding helping skills with each step. The underlying attitudes are addressed in Part III of the skill development section, the five-step communication model is presented in Part IV, and the corresponding helping skills are discussed in Part V.

Relationship Goals

The segment on relationship goals addresses goals for the helping relationship, from initial rapport to termination of the helping relationship. Helping relationship goals are addressed in Part V of the skill development section.

❖ EXERCISE 14:   STAGE STRUCTURE FOR THE HELPING PROCESS
RELATIVE TO CURRENT PERSONAL APPROACH

How does your current approach to the helping process from Exercise 13 fit with the Stage Structure of the Helping Process?

Helper Personal Growth

To learn to fly, we go to a pilot for lessons; to learn to weave, we go to a weaver. To learn, we seek out those who demonstrate the skills we desire.

❖ **EXERCISE 15:   THE COUNSELOR AS CLIENT**

Respond to the statement "You can't help others go beyond where you yourself have gone." What correlation is there between your own growth and your ability to facilitate others' growth?

The purpose of the helping relationship is to facilitate human growth and development. As humans, both helper and client are on their own respective journeys of growth and development. To suggest that you are capable of facilitating the growth of a fellow human being implies inherently that you are growing yourself because you, too, are a human being. If you are *not* involved in your own growth, how can you effectively help someone else do what you are not able to do yourself?

❖ EXERCISE 16:   THE IMPORTANCE OF HELPER GROWTH IN
EFFECTIVE HELPING

What effect does a helper's personal growth have on his or her ability to be an effective helper?

The importance of helpers' personal growth to their ability as effective helpers cannot be overemphasized. Three areas specific to this dimension include the parallel process of growth facilitation, the importance of boundary distinction, and the helper as a role model.

❖ EXERCISE 17:   PARALLEL PROCESS OF GROWTH FACILITATION

Identify the people you tend to seek out when you are distressed and/or need help.

How many of the people you named are people you perceive as incapable of handling a similar situation for themselves?

Growth facilitation is a parallel process; if you are not involved in your own growth, it's less likely you'll be able to help others with theirs. The section on self-development in this manual acknowledges the importance of helper personal growth.

Boundary distinction refers to your ability to differentiate clearly between yourself and others, distinguishing personal issues from others' issues. Boundary distinction includes self-awareness, appropriate responsibility-taking, and identification of potential counselor-client relationship issues.

❖ EXERCISE 18:   BOUNDARY DISTINCTION

How clear are the boundaries in the examples below? Identify the extent to which each reflects clear boundary distinction. Justify your responses.

1. CLIENT: *I just don't know what to do. My car broke down getting here, and now I don't even know how I'm going to get home.*

    COUNSELOR: *Oh, that's too bad. Actually, you're my last client, so I could give you a ride.*

    How clear is this boundary?

2. CLIENT: *I don't know what I'd do without our sessions. Whenever I think about ending therapy I go into a cold sweat! I think I'll just plan to keep coming forever.*

    COUNSELOR: *Well I'm not planning to go anywhere.*

How clear is this boundary?

3. CLIENT: *I'm in a total panic. My car's broken down in the parking lot and I don't even know how I'm going to get home.*

   COUNSELOR: *Sounds like you're in a tough situation. I'm wondering if you've begun thinking about your options.*

How clear is this boundary?

Role modeling is one of our most potent learning processes. As a helper, you are in a powerful position for role modeling. You will role model—whether you model good or poor psychological health.

❖ EXERCISE 19: ROLE MODELING

Who have been your most important role models? Have you learned anything from them that you could not have learned any other way?

A Stage Structure for the Helping Process: Pre-Stage Attitudes of Rights, Respect, and Responsibility

Attitudes speak louder than words. Words cannot substitute for attitudes. We cannot develop effective helping skills, therefore, without addressing attitudes.

CHAPTER 8

Beliefs Affect Attitudes

What we believe for ourselves affects our attitude toward others. For example, if you value honesty, you will likely value honesty not only in yourself but also in others. If you place importance in honoring commitments and doing what you say you will do, you probably admire this trait in others. Your attitude, then, will likely be affected by that belief when someone is dishonest or breaks a promise. Our beliefs affect our attitudes.

❖ **EXERCISE 20:** **BELIEFS AFFECT ATTITUDES**

Think of a person whom you dislike. Describe the personality, behaviors, and attitudes the person exhibits.

Now identify something about that person that you wish to be more like (e.g., what does that person have permission to do or be that you don't give yourself permission to do or be)? In other words, what beliefs that you hold does this person not live by?

33

ATTITUDES SPEAK LOUDER THAN WORDS

My client walked in, face downcast, and dragged himself over to the chair. He fell into it and slouched down. Without looking up, he sighed heavily. When asked how he was doing, he responded slowly in barely a whisper, "Oh, I'm doing just great. Best I've ever been."

Attitude, which is communicated nonverbally, speaks louder than our words. This client was clearly not doing great, despite his words to the contrary. When attitudes and words contradict each other, attitude is typically the more accurate message.

❖ **EXERCISE 21:**   **ATTITUDES OVERRIDE OUR WORDS**

Ask a partner to make five or six statements in which his or her attitude says the opposite of the words. For example, "I'm really upset" (said lightheartedly, while smiling). Try responding just to the words. Reverse roles with your partner. Discuss the effect of attitude on words. Which seems a more accurate message? How often do you say one thing with your words and mean another with your attitude?

ATTITUDES AFFECT BEHAVIOR

"Say you're sorry." "I'm sorry," the ten-year-old growled. "Now give your sister back her crayons." The crayons came rocketing through the air, scattering as they hit the floor. Attitudes, reflecting our underlying beliefs, affect not only our words but also our behavior. In this case, the ten-year-old apparently believed he deserved to have the crayons. Although he gave them back, his attitude communicated his belief that he shouldn't have to.

❖ **EXERCISE 22:   ATTITUDES AFFECT OUR BEHAVIOR**

With a partner, try to communicate three or four different attitudes through your behavior. For example, sitting with your arms crossed, glaring at your feet, communicates an attitude of defiance. Try to communicate your attitude as accurately as you can through your behavior.

ATTITUDES AFFECT OUR SKILLS

Tom, 27, had been teaching math in the public high school for five years when his younger brother Aaron, 18, was arrested for dealing drugs. Tom blamed Aaron's trouble on his rowdy friends, whom Tom accused of being a bad influence. Today one of Tom's students came up to him after class. The boy's hair was shaved around the bottom of his head, with a pony tail left on top. He was wearing a leather jacket and had an earring in his ear. Although he wasn't one of Aaron's friends, he looked a lot like them. "What do you want?" Tom almost snarled.

Because attitudes affect our words and behaviors, they affect our ability to use helping skills. Our beliefs and attitudes go with us into our helping relationships. It is essential, therefore, that we consider our attitudes if we are to develop effective helping skills.

CHAPTER 9

Attitudes of Rights, Respect, and Responsibility

The client's name is Debbie. She is 21, is receiving welfare, and has three children ages 5, 3, and 1½. Each child has a different father. Debbie is currently in an on-again, off-again relationship with the father of the youngest. He is a 16-year-old high school dropout. Both he and Debbie are unemployed. Two weeks ago, Debbie's oldest child set fire to the crib where the baby was sleeping. The child protective services became involved, and Debbie now has a choice of getting counseling or losing her children, who are currently in protective custody. Debbie is grudgingly coming in to see you.

❖ **EXERCISE 23:   ATTITUDE AWARENESS**

What are your reactions to Debbie? What attitudes were you aware of as you read about her situation? Be as honest as you can in identifying positive and negative reactions.

We all have beliefs and values. It is impossible for us to be value free. Our beliefs go with us into the helping relationship. Although we can't be value free as we interact with our clients, it *is* possible for us to make conscious decisions about the imposition of our values and beliefs on our clients. We must first, however, be aware of our underlying beliefs and attitudes.

❖ **EXERCISE 24:** **UNDERLYING BELIEFS AFFECTING ATTITUDE**

Identify the underlying beliefs that might affect your reaction and attitude toward Debbie and subsequently affect your ability to be an effective helper with her. An example of such a belief is that people shouldn't have children if they can't support them.

❖ **EXERCISE 25:** **EFFECT OF ATTITUDE ON HELPING**

Form a triad. Have one person role play Debbie. In the role play, Debbie will be reluctant to enter counseling, appearing to have a chip on her shoulder. You take the role of helper and respond to Debbie. The third person should observe and take notes.

Describe your approach. What attitudes were you aware of? How were they affecting your approach? Assess your effectiveness.

Our beliefs regarding an individual's right to be who he or she is (rights), respect for an individual's capability and inherent worth as a human being (respect), and an individual's responsibility for his or her actions and choices (responsibility) influence our attitudes when we are helping others. Attitudes are a powerful form of communication that can counteract and override our words and behavior. Attitudes, therefore, are the foundation for verbal and behavioral helping skills. The following conceptual attitudes are proposed as a pre-stage foundation for the helping process (see Table 6.1). They are hierarchical. Each is an extension of and builds on the previous one.

1. *Rights.* Individuals have a right to be who they are—a right to their own feelings, beliefs, opinions, and choices (short of harm to others).

2. *Respect.* Individuals are capable, able, and inherently worthy of unconditional acceptance as human beings. If I believe you have a right to be who you are, I need to accept and respect who you are.

3. *Responsibility.* Individuals are ultimately responsible for their own actions and choices, and for making decisions regarding who they are and how they choose to live their lives. If I believe you have a right to be who you are, and respect that right, I must recognize that only you know who you are, and I must, therefore let you take responsibility for making decisions consistent with who you are.

While we may agree with the validity of these attitudes in theory, often they are challenging to implement in practice. Rather than communicating to our clients an attitude of acknowledgment of their *rights*, we attempt to control. While we say we *respect* their uniqueness, we still find ourselves feeling judgmental. Instead of allowing them to take *responsibility* for their life choices, we find ourselves trying to rescue or blame them. We sometimes forget the metaphor that, as helpers, we can lead a horse to water but the horse alone can drink.

The following chapters address the challenge of implementing an attitude that demonstrates acceptance of a person's *rights*, that accords that person *respect*, and that expects appropriate *responsibility-taking* from that person.

CHAPTER 10

Rights versus Control

When your attitude reflects your belief that individuals have the right to be true to themselves, a right to their own feelings, beliefs, actions, and choices (short of harm to others), you are demonstrating unconditional acceptance of individuals as human beings, just as they are. We've probably all heard the saying, "You can't love someone else until you can love yourself." This maxim is true because our beliefs about ourselves affect our attitudes toward others. We can't give to others something we don't have. If we don't believe we have the right to be true to ourselves, we are not likely to accord this right to others. If we don't believe we have the right to our feelings, opinions, and choices, we probably do not believe others have this right, either.

What we give to ourselves, we give to others.

(WILLIAMSON, 1992, P. 33)

❖ EXERCISE 26: ATTITUDE OF RIGHTS

The following clients are coming in to see you:

- Joe, 22, is an African American who lives with his girlfriend Greta, 29. Greta is originally from Germany, is pregnant, and wants Joe to move there with her.

- Linda, 31, is an Anglo woman who is struggling to make a living as a psychic, is in a relationship with Sharon, 19, and wants Sharon to move in with her.

- Maria, 40, is a Hispanic woman with five children. A devout Catholic, she doesn't want more children but is afraid to use birth control.

Assess your attitude toward the rights of each client.

❖ **EXERCISE 27:**   **ACKNOWLEDGING INDIVIDUAL RIGHTS**

Do a role play in which you are the counselor and you're seeing three different clients. Have the clients represent a diversity of cultural, ethnic, spiritual, and educational backgrounds, as well as sexual orientations and ability/disability. Who decides whether each is satisfied with his or her life? Assess your ability to respond with an attitude that is respectful of each client's rights.

RIGHTS VERSUS IMPOSING CONTROL

Belief in people's rights means that you think individuals have the right to be themselves and to make their own choices regarding how they live their own lives. To the extent that we really believe individuals (including ourselves) have the right to be true to themselves, we do not try to impose external control.

Internal control means you are in control of your own beliefs, choices, and actions. *External control* means others control your beliefs, choices, and actions. While many people *say* they believe in individuals' rights to be themselves, their *behavior* often reflects an attempt to impose external control. For example, this statement—"I support your doing what you believe is right for you; I may not agree, but you know what's best for you"—reflects an acknowledgment of individual rights and support of *internal control*. This statement—"You should apologize to your father; even if he was wrong, you should take the initiative to clear this up"—imposes *external control*.

Imposing control reflects an underlying belief that you know what's best for the other person. It suggests that you know what that individual needs better than he or she does.

❖ **EXERCISE 28:   ACKNOWLEDGING RIGHTS VERSUS IMPOSING CONTROL**

Read the following excerpts. Identify the extent to which they represent an acknowledgment of individual rights or reflect an attempt to impose control.

"So, trying to go to school and work full time is really getting to you, huh? That is a lot to try to do. And you're thinking dropping out of school is the answer. Well, you're in a tough position. I bet it's hard to know what to do."

"So, trying to go to school and work full time is really getting to you, huh? Well, dropping out of school certainly isn't the answer. Don't do that; it will just make the situation worse!"

"So, trying to go to school and work full time is really getting to you, huh? Well, you don't have it half as bad as you think you do. You better just hang in there and make the best of it."

DIRECT VERSUS INDIRECT MANIPULATION

In attempting to control another, we can use a direct or an indirect approach. An example of direct manipulation is this: "If you don't go with me to visit my mother, our relationship is over!" Indirect control generally includes an unstated but implicit message; for example, "Well, I'd think about the consequences of your decision, before deciding not to go, if I were you."

❖ **Exercise 29:   Manipulation**

Describe the most recent situation in which you attempted to control someone through direct manipulation.

Describe the most recent attempt you made to control someone through indirect manipulation.

Respect versus Judgmentalness

An attitude of respect reflects your belief that individuals are capable, able, and inherently worthy of acceptance as human beings. Inherent in respect is acknowledgment and honoring of individual diversity in culture, ethnicity, spirituality, sexual orientation, family, educational and socioeconomic background, and ability/disability. Respect shows acknowledgment and acceptance of individuals' capability and rights to make their own decisions and choices regarding their own lives. This applies to you as well as to others.

Judgmentalness results when one individual imposes his or her values, beliefs, standards, or opinions on another. For example, "You shouldn't go back to work until your kids are in school," or "You should finish school before you get married" are opinions. You become judgmental when you impose them on others with the message that unless they comply or agree, you will think less of them. You become judgmental when you label your opinion "right" and another's opinion "wrong."

Judgmentalness becomes a force of habit, a way of responding—to yourself, others, and circumstances. When you are being judgmental, you automatically judge the situation as "good" or "bad," "right" or "wrong." In reality, whether a situation is "good" or "bad" is a matter of perspective, and this is sometimes difficult to know. Consider the following old folk tale, for example:

An old man and his son worked a small farm, with only one horse to pull the plow. One day, the horse ran away. "How terrible," sympathized the neighbors. "What bad luck." "Who knows whether it is bad luck or good luck," the farmer replied. A week later, the horse returned from the mountains, leading five wild mares into the barn. "What wonderful luck!" said the neighbors. "Good? Bad? Who knows?" answered the old man. The next day, the son, trying to tame one of the horses, fell and broke his leg. "How terrible. What bad luck!" "Bad luck? Good luck?" The army came to all the farms to take the young men for war. The farmer's son was of no use to them, so he was spared. "Good? Bad?"—(SOURCE UNKNOWN)

ATTITUDE REFLECTING RESPECT VERSUS JUDGMENTALNESS

An attitude of respect means responding (to self, others, and circumstances) with acknowledgment, understanding, and acceptance. An attitude of judgmentalness means responding with a determination of good, bad, right, or wrong. Saying a belief is "good" and "right" is just as judgmental as calling it "bad" and "wrong."

❖ **EXERCISE 30: RESPECTFULNESS VERSUS JUDGMENTALNESS**

Read the student's statement below. To what extent does each of the following responses reflect respect or judgmentalness?

STUDENT: *"I've decided to drop chemistry. I just can't keep up with the lab assignments and it's interfering with my work hours, not to mention my relationship with my girlfriend. She gets really upset when I don't have time to spend with her."*

Responses:

"Good. That sounds like a good idea."

"So you think dropping chemistry would resolve your time problems."

"I think you're fooling yourself. Dropping chemistry's not going to help. The time will just get taken up with something else."

USING RESPECT

Judging is not the same as judgmentalness. We are always making judgments. Judgments are assessments of circumstances from the viewer's perspective. Judgmentalness is a position of deciding that something is "good" or "bad" and the imposition of that decision on someone or something else. In using respect, we first make an assessment, and second, make a judgment, based on our goal, about how to respond with respect in facilitating achievement of that goal. Imagine that a friend or client of yours is distraught and considering getting a divorce. Your response might be based on the following:

1. *Assessment*: He's upset. It's probably hard for him to think clearly right now. He may be reacting.
2. *Goal*: To acknowledge and accept the reality of his circumstance. To be understanding of his experience.
3. *Response to Facilitate Achievement of Goal*: "It really sounds like something has happened to trigger a lot of feelings for you—even to the point of considering some pretty strong reactions."

❖ **EXERCISE 31:   SHOWING RESPECT**

Describe a circumstance to which you recently responded. Use the three steps above:

Event Description:

Assessment:

Goal:

Response:

Did your goal and response demonstrate respect? If not, how might you re-identify your goal and respond with respect?

CHAPTER 12

Responsibility versus Rescuing or Blaming

When you acknowledge individuals' responsibility, you are reflecting your belief that people have the right to, and therefore responsibility for, their own actions and choices. They have a responsibility to make decisions consistent with their own beliefs as these affect their own lives.

If we believe that we all have the right to be true to ourselves and we respect our capability to make choices consistent with being true to ourselves, then we will logically assume that we all must take responsibility for making our own choices.

RESCUING VERSUS RESPONSIBILITY

Rescuing, for our purposes, refers to the voluntary, *unnecessary* assumption of responsibility for another person's feelings, choices, or actions. Rescuing is implicitly disrespectful of the other person's ability to take responsibility for himself or herself.

❖ **EXERCISE 32: RESCUING VERSUS RESPONSIBILITY**

In responding to the following client statement, indicate the extent to which each response reflects appropriate responsibility-taking versus rescuing:

CLIENT: *I just don't know what to do. My father is furious with me for choosing theater arts rather than pre-med. Now he's upset because I'm going out with someone outside our religion. He's threatening to cut off all financial support, and I don't think I could make it without his help.*

RESPONSE 1: *It sounds as though you're concerned about your father's reaction to your decisions primarily because his financial support is important to you.*

RESPONSE 2: *If you can't make it without your father's financial support, then perhaps you'd better start making some decisions that will keep him happy.*

RESPONSE 3: *Well, you know what you can do? You can apply for financial aid. In fact, I've got some application forms right here.*

49

❖ EXERCISE 33:   RESCUING—SELF-ASSESSMENT

Describe a time when you voluntarily and unnecessarily tried to rescue someone.

Rescuing reflects the *rescuer's* needs. Rescuing is typically motivated by needs such as a lack of confidence in the capability of the person being rescued, a need to feel important, or a need to be needed.

❖ EXERCISE 34:   RESCUING— SELF-MOTIVATION

What needs motivated you to rescue in the above example?

BLAMING VERSUS RESPONSIBILITY

Blaming is imposing judgment regarding a specific incident. Blaming often reflects either a lack of appropriate responsibility-taking for oneself or a botched effort to rescue someone else.

❖ **EXERCISE 35:   BLAMING VERSUS RESPONSIBILITY**

In the following examples what evidence do you see of a lack of appropriate self-responsibility? Botched efforts to rescue someone else?

"I can't believe that test! There weren't any questions that had been included in the review! And he didn't tell us to study the stuff in that last section. This guy is a terrible teacher."

"I can't believe you! How many times did I tell you that you needed to study for that test? I told you it was going to be hard. I tried to get you to come study with me—so don't complain to me that you didn't do well!"

❖ **EXERCISE 36:**

Describe the most recent time you blamed someone else rather than taking responsibility for your choices, feelings, or actions.

Can you think of a time you judged and blamed someone else, after that person didn't allow you to rescue him or her? What's your most recent example?

A Stage Structure
for the Helping Process:
A Five-Step Communication Model

Communication skills are the core of the helping process.

CHAPTER 13

Communicating to Facilitate Self-Understanding

We've identified the goal of the helping relationship as growth toward increased psychological health as evidenced by a positive self-concept. Correspondingly, the goal of communicating within a helping relationship is the facilitation of self-understanding. Self-understanding is part of growth toward a positive self-concept.

We've identified our purpose and goals, answering the question "What do we want to accomplish?" as the facilitation of growth toward psychological health and a positive self-concept (see Part I). We've clarified our philosophy of human growth and development, in answer to the question "What is growth and psychological health?" (see Part II). We've identified a stage structure for the helping process, in answer to the question "How do we go about facilitating growth toward a positive self-concept?"

The stage structure for the helping process (see Table 6.1) addresses helper skills for facilitating growth. Helper skills include underlying beliefs and attitudes (see Part III), a communication model, and facilitative skills. We are ready, in Part IV, to address communication and the Five-Step communication model.

GOALS OF COMMUNICATION

The three basic goals for communicating are (1) to have your perspective understood, (2) to understand another's perspective, and (3) to help someone understand his or her own perspective. Lecturing, conveying information, or sharing your point of view are examples of goal 1: to have your perspective understood. Listening to a lecture, listening to directions, or listening to another's point of view are examples of goal 2: to understand another's perspective. Empathizing with a partner, paraphrasing what a child has said so he or she can hear it, or double-checking a friend's stated rationale for a major decision are examples of goal 3: to help others understand their own perspectives. While all three goals are useful at different times, goal 3 best corresponds with our goal in the helping relationship: to help others better understand themselves and their own perspectives—to facilitate self-understanding.

❖ EXERCISE 37:   GOALS OF COMMUNICATING

Form a dyad and carry on a conversation. First, try focusing on goal 1, then goal 2, and finally, goal 3. Which felt most familiar? What effect did each have on the interaction? How did each feel to your partner?

PATTERNS OF COMMUNICATING

Because we have been communicating for many years, we have developed patterns of communication. We form habits, which we continue, generally unconsciously. We continue our communication patterns, regardless of how effective they are, until we are confronted by a situation in which our communication skills fail us, or until we decide to do a conscious self-analysis. The self-analysis can include the following criteria:

1. *Data Collection.* Collect information (as objectively as possible) about your patterns. You might set up a triad in which you are communicating with another person while the third person acts as an observer and notes patterns.

2. *Pattern Identification.* Identify patterns based on input from your observer, your communication partner, and yourself. Patterns might be asking questions, finishing sentences for the other, giving advice.

3. *Effectiveness Assessment.* Analyze the effectiveness of your communication patterns based on the *goal* of the communication process—for example, advice giving didn't help facilitate self-understanding.

❖ EXERCISE 38:   COMMUNICATION PATTERNS

Form a triad. Choose one person to be the client and the other person to be the observer. Have the client share something of personal concern. Respond with communication goal 3 to facilitate self-understanding. After 10–15 minutes, solicit feedback and do a self-assessment of your patterns of communicating.

Data Collection:

Pattern Identification:

Effectiveness Assessment:

CHAPTER 14

A Five-Step Communication Model

The following five-step communication model can be used effectively with all three basic goals of communicating. Our purpose here, however, is to apply it to our identified goal of facilitating growth and self-understanding.

COMMUNICATION MODEL EMBEDDED IN STAGE STRUCTURE FOR THE HELPING PROCESS

The five-step communication model is embedded in the helper skills segment of the stage structure for the helping process (see Table 6.1). Listening, centering, and empathizing, steps 1, 2, and 3, are used to facilitate the goal of Stage 1: understanding the client's presenting problem. Step 4, focusing, is used to facilitate the goal of Stage 2: understanding the client's underlying issues. Providing directional support, step 5, is used to facilitate the goal of Stage 3: direction and change.

PHILOSOPHICAL BASIS IN MARTIAL ARTS

The model is based on a martial arts philosophy of bringing the energies between yourself and others into harmony. The philosophy is based on a principle of alignment that, when used effectively, prevents conflict and promotes understanding.

When you use the philosophy of martial arts in a potential physical confrontation, such as avoiding a conflict with an attacker coming toward you, you apply the following five steps:

1. *Observe:* Pay attention; you need to know what your opponent is doing.
2. *Balance:* Balance yourself physically, with your weight over your feet.
3. *Align:* Go *with* the momentum of the aggressor rather than trying to block him or her.

4. *Direct:* Once you're in alignment, *then* go with the opponent and physically direct the momentum and direction of his or her body.

5. *Follow through:* Let your body's energy follow through, just like a batter continues his swing after hitting a baseball.

Using the martial arts principle of alignment, you have facilitated direction while maintaining the philosophy of harmony.

MARTIAL ARTS APPLIED TO COMMUNICATION

The five steps of the martial arts correspond to basic principles of effective communication. The principle of paying attention and observing in martial arts is central to *listening* in communication. Balancing physically corresponds to *centering* emotionally in communication. Aligning in martial arts, which reflects going with the physical momentum of the opponent, is equivalent to *empathizing* with a client's perspective and presenting problem. Going with the opponent physically in order to assist with his or her momentum and direction corresponds with assisting our clients in *focusing* on their underlying issues. Finally, following through in martial arts is equivalent physically to what we do in providing *directional support* for our clients as they attempt to implement new direction and change in their lives.

Martial Arts		Communication Model
1. Observe	⇨	Listen
2. Balance	⇨	Center
3. Align	⇨	Empathize
4. Direct	⇨	Focus
5. Follow Through	⇨	Provide Directional Support

The three parts of the stage structure (presenting problem, underlying issues, and direction and change) are hierarchical (see Table 6.1), but you may move back and forth between them as new issues emerge. Similarly, the five steps of the communication model are hierarchical and cumulative. In other words, you don't listen only once and then move on; you keep listening as you add subsequent steps.

The five steps correspond to the stage structure's three stages (see Table 6.1) *and* they can be applied to individual sessions and individual interactions as well. To apply the five steps, we need to develop each of the respective communication skills: listening, centering, empathizing, focusing, and giving directional support.

COMMUNICATION SKILLS OF THE
FIVE-STEP COMMUNICATION MODEL

Listening

Listening is a prerequisite to effective responding; however, for most of us, it is the weakest, least well-developed communication skill.

> LISTEN
>
> When I ask you to listen to me and you start
> giving advice,
> You have not done what I asked.
>
> When I ask you to listen to me and you begin
> to tell me why I shouldn't
> feel that way,
> You are trampling on my feelings.
>
> When I ask you to listen to me and you feel
> you have to do something
> to solve my problems,
> You have failed me, strange as that may seem.
>
> Listen! All I asked was that you listen, not
> to talk or do
> —just hear me.
>
> Advice is cheap:
> twenty-five cents will get you both Dear Abby
> and Billy Graham
> in the same newspaper . . .
>
> . . . So, please listen and just hear me. And,
> if you want to talk, wait a minute
> for your turn; and I'll listen to you.
>
> —ANONYMOUS

❖ **EXERCISE 39: GRAPEVINE**

Have 8–15 persons stand in a circle or line. Have the first person write down on a piece of paper five or six sentences of information and then whisper the words once only in the ear of the second person. Make sure no one else can hear. The second person whispers what he or she heard, as accurately and completely as possible, in the ear of the third person, and so on, until the last person is told the information. Have the last person repeat aloud what he or she heard. Have the first person verify for accuracy. What does this tell you about our ability to listen? How well are you able to listen?

Effective listening involves paying attention, hearing, understanding, and remembering what you've heard. Specifically, listening can be broken down into these steps:

1. Attending (paying attention)
2. Hearing
3. Explicit comprehension (understanding the content)
4. Implicit comprehension (understanding the nonverbal or implied message)
5. Understanding (accurate understanding of the combined verbal and nonverbal message)
6. Remembering

❖ **Exercise 40: Listening**

In a triad, have one person share about two or three minutes' worth of information regarding a concern or frustration. Have a second person observe and take notes. Respond with your understanding of what the first person shared with you as accurately as you can. Solicit feedback regarding accuracy. Do a self-analysis of your ability to listen (steps 1–6).

Attending (paying attention):

Hearing:

Explicit comprehension (understanding the content):

Implicit comprehension (understanding the nonverbal or implied message):

Understanding (accurate understanding of the combined verbal and nonverbal messages):

Remembering:

Centering

The student counselor walked into the session having just discovered he got a "C" on his theories of counseling exam. He was feeling depressed and disappointed. In opening the session, his client, who is trying to take more responsibility for her decisions, began, "I'm not sure I'm moving fast enough . . ." Before she could finish her thought, the student bristled, "Well, I'm doing my part. I know what I'm doing." The client was stunned. "I wasn't going to say you aren't, or don't," she said. "I was going to say that I think I'm procrastinating."

Centering means being in balance with your thoughts, feelings, and behaviors. It means being congruent and aware of what's going on with you. Centering requires listening to yourself, and allows you to recognize and separate your issues from the issues of the person with whom you are talking. The student counselor above was off center. He wasn't differentiating his issues from his client's. If you are not centered, you are more likely to have your "buttons pushed," to take things personally, and to react defensively. To develop our ability to center, we must first be aware of our "buttons" that are particularly sensitive, such as being told what to do, being criticized, or receiving anger. Second, we can work toward desensitizing them.

❖ **EXERCISE 41: BUTTON DESENSITIZATION**

Choose a partner with whom to work on "desensitizing your buttons." Sit directly across from your partner with an open body posture. The objective of the exercise is for you to stay centered. Staying centered in this exercise will be demonstrated by your doing the following:

1. *Make a clear, firm, neutral statement to your partner to "Raise your hand," then "Lower your hand."*

2. *Make the statement each time as though it is the first time you've said it.*

3. *Avoid use of nonverbal behaviors that imply pleading or threatening, such as smiling, grimacing, or laughing.*

4. *Keep eye contact and an "openness" to your partner (e.g., do not go into a staring, robot-like trance, concentrating on not being impacted by your partner). Do not ignore your partner!*

Your partner, on the other hand, will be trying to "push your buttons." He or she will start out being cooperative with your request (raising his or her hand

and lowering it) but then will begin using a variety of responses to push your buttons. Responses might include anger ("Can't you make up your mind?"), seductiveness ("Where would you like my hand?"), whining ("But I'm so-o-o tired!"), uncooperativeness ("Forget it!"), sarcasm ("Like I really want to raise my hand"), hostility ("Just try telling me once more!"), defiance (Not doing it), and so on.

First, your partner should use a range of responses to see which ones trigger buttons. Then he or she should focus on one, increasing its intensity as the button gets desensitized.

During the process, whenever your partner sees that your button is being pushed (any reaction other than a calm, collected, "centered" response, such as laughing, smiling, staring, looking away), he or she says "Stop," lets you collect yourself and, when you are ready, starts the process again by saying "Begin."

Following the exercise, do a self-analysis, including the following:

What buttons do you have that are particularly sensitive (e.g., when others get angry or whine)?

How does it feel when you get off center?

How does being off center affect your ability to communicate?

Empathizing

As I watched through the one-way glass, I listened to the client describe his experience for my practicum student. "I didn't have any choice," he said. "If I stayed I was going to lose my sanity. All I could think to do was pack up and leave. I couldn't bear to say 'good-bye' so I just left a note. I know my wife's confused. I know it's hard on my kids— especially my four-year-old; she won't understand where her dad disappeared to . . ." The practicum student responded with empathy, "It sounds like you found yourself in a difficult position and felt you had to make a decision to leave to maintain your 'sanity.' It also sounds like you're concerned about the effect your leaving had on your wife and kids."

Empathizing is looking at the situation from other people's perspectives, and communicating your understanding of what you see. It means looking at the world through their eyes and understanding their experience—their thoughts, feelings, and behaviors. Empathizing is *not* the same thing as agreeing, condoning, or liking. It simply means *understanding*.

In the situation above, the practicum student had just experienced having her husband leave her with a young child. She held very strong beliefs about marriage and staying together. She did not, therefore, agree with, condone, or like what her client was saying.

The point is that she didn't have to agree, condone, or like what he said in order to empathize. Empathizing means understanding the client's perspective, and then communicating that understanding without judging it.

❖ **EXERCISE 42:   EMPATHIZING**

Read the following statement. Respond by writing a statement of empathy below.

"I feel terrible. All of a sudden it looks like I'm not going to get the job I interviewed for, just at a time when my husband's really preoccupied with his work. That just makes me feel more disappointed, and I wonder if I should just give up trying to get a job."

What's your understanding of this person's perspective?

There are several specific facilitative skills that use empathy, such as explicit empathy, implicit empathy, and confrontation. Empathy is addressed further in Part V.

Focusing

"I just can't manage in this marriage anymore," the client said emphatically. "I don't know what happened, but it's just not working. My mother doesn't think it is, either. Her marriage wasn't any better—in fact, it was a mess. Dad left when I was nine. I guess that's why I married young." "Hmm," the student counselor nodded. "I have a lot of trouble with my two-year-old and have a hard time sleeping."

Focusing means directing attention to a point or an issue to help achieve greater clarity. When there is a lack of focus, as shown by the counselor's response in the example above, it creates a sense of purposelessness in the counseling session.

The central goal of the helping process is to facilitate growth toward a positive self-concept through client self-understanding. The process of understanding involves, in part, clarifying issues and bringing them into focus. Facilitating self-understanding is like putting a jigsaw puzzle together. You make an initial identification of what you think the picture looks like, you get all the pieces, turn them over, and put together the edge pieces to provide an outline. You help your client focus on each piece, looking for similarities of colors, shapes, and images. The goal is to fit pieces together so that a clear image will come into focus. Both client and counselor can initiate focusing. Only the client, however, can decide whether the resulting image is accurate.

Focusing, then, is central to the helping process. It is the intentional directing by the helper of attention toward specific aspects of client communication, for the purpose of facilitating clarification, insight, and understanding. Just as in focusing with a camera, it refers to both directing attention to significant points and clarifying the image, concept, or insight.

In the martial arts, focusing is a directing of *energy toward*. In communication, focusing is directing *attention to*. It is directing attention in the interaction so that meaning and understanding become clearer and clearer. Obviously, both client and counselor can initiate focusing. In terms of the *skill* of focusing, however, our concern is with the counselor.

Focusing can be done using a variety of skills that reflect empathy or expression. An empathic statement is one that verbalizes your understanding of the *other person's* perspective; an expressive statement verbalizes *your* perspective. Both empathic and expressive perspectives can provide focus. For example, "It sounds as though (from what *you're* saying) you're not feeling like attending the wedding" is an empathic perspective. "I don't think (from *my* viewpoint) you're well enough to attend the wedding" is an expressive perspective.

❖ EXERCISE 43: FOCUSING

Respond to the following statement, using an empathic and an expressive focus:

"What am I going to do? Now that my sister's married and moved, there's no one to take care of my mother except me. She can't live by herself anymore, but I don't have room for her to live with us. We don't even get along very well."

Empathic focus:

Expressive focus:

We can use a variety of facilitative skills to promote focusing, such as implicit empathy, confrontation, self-disclosure, and immediacy. Specific skills are addressed in Part V.

Providing Directional Support

Providing directional support is providing help and support as the client determines direction, makes decisions, and implements desired changes. The focus increasingly centers on action and change.

For example:

> CLIENT: *I know now that I want to be more assertive, and share what I think and how I feel. I just don't know how to do it. I'm so used to just not saying anything.*
>
> HELPER: *It does sound as though you've become clear that sharing more of how you think and feel is something you really want to do. Perhaps we can brainstorm some ideas for how you might be able to do that.*

❖ EXERCISE 44:   DIRECTIONAL SUPPORT

To the following statement, respond using directional support:

"I know I need to start studying more, and more regularly. I just have a hard time doing it. There always seems to be something to distract me."

We can use a variety of facilitative skills to promote directional support—directionality, implementational support. These specific skills are addressed in Part V.

A Stage Structure for the Helping Process: Stages and Skills

Facilitative skills are specific helping skills used for the purpose of promoting growth. A usable structure for the helping process needs to include a systematic, operational organization of facilitative skills.

CHAPTER 15

Stage Structure Overview

As I watched through the one-way glass, I saw that the counselor did not noticeably move in over 20 minutes. She sat completely still in her chair with her feet flat on the floor and one hand on each knee. She looked straight ahead at her client. After 20 minutes she appeared stiff and self-conscious. Later, in supervision, she explained that she was trying to use the skill of attending behavior. "To attend" she explained, "the book said to keep an open posture and face your client squarely."

This is an example of a counselor who was so distracted by trying to *do* the behavior of a particular skill that she lost sight of why she was doing it.

PURPOSE OF STAGE STRUCTURE AND SKILLS

Helping skills are tools used to facilitate a goal. They are not a goal in themselves. The skill of attending, for example, (see the helper skills segment of Table 6.1) is used to communicate that you are there for your client and ready to listen. Ironically, if you are too preoccupied trying to *do* the specific behavior associated with the skill—keep an open body posture—you can lose sight of your goal, and end up *not* being there for your client!

The same principle is true when we consider the purpose of a stage structure for the helping process. The purpose is to provide an organized approach to the helping process in order to facilitate client growth and self-understanding. The structure identifies a stage hierarchy, corresponding steps of communication to facilitate stage goals, and potential specific helping skills (see Table 6.1). The reason for identifying specific skills is to provide guidance in facilitating growth. Your goal is not simply to use the skills. The skills are merely a means to facilitating the goal. Your goal is to facilitate growth.

FACILITATIVE SKILLS IN THE HELPING STRUCTURE

The stage structure for the helping process identifies three hierarchical stages for the facilitation of growth. Each stage includes one or more of the five communication model steps as well as specific skills to guide the helper in facilitating client growth (see Table 15.1).

STAGE FOCUS	HELPER SKILLS	
Pre-Stage Assumptions: Beliefs and Attitudes	Attitudinal Goals: Rights Respect Responsibility	
Stage 1: Presenting Problem	Communication Model 1. Listen 2. Center 3. Empathize	Facilitative Skills Attending Genuineness Positive Regard Boundary Distinction Explicit Empathy Concreteness

Stage 2:
Underlying Issues
	4. Focus	Empathic:
		  Implicit Empathy
		  Confrontation
		Expressive:
		  Self-Disclosure
		  Immediacy

Stage 3: Direction and Change	5. Provide Directional Support	Directionality Implementational Support

TABLE 15.1
Helper Skills

Steps one, two, and three of the communication model—listening, centering, and empathizing—are used to facilitate the goal of Stage 1—clarifying the presenting problem. In Stage 2, step 4 of the communication model, focusing is added. Step five of the communication model, directional support, is added in Stage 3. Specific facilitative skills—attending, genuineness, explicit empathy—are used to help accomplish each respective communication step—listen, center, and empathize (see Table 15.1).

CHAPTER 16

Stage 1 Helping Skills

In Stage 1 of the helping process, the goal is to help clients clarify their presenting problem. Clarification is promoted by helping clients tell their stories, identifying initial awareness of thoughts, feelings, and behaviors, and identifying what clients view as their initial problem. The goal of the helping relationship is the development of rapport and trust (see Table 6.1).

Helper skills that can be used to promote clarification of the presenting problem and the development of rapport and trust include steps one, two, and three of the communication model that were addressed in Part IV, and their corresponding facilitative skills that are addressed in this chapter.

Communication Model Step		Facilitative Skill
1.	Listen	Attending
2.	Center	Genuineness Positive Regard Boundary Distinction
3.	Empathize	Explicit Empathy Concreteness

LISTENING

The first step of the communication model is listening. A specific facilitative skill that can facilitate the listening process is attending.

Attending

Attending is both the physical process of being alert and ready to receive communication, and the mental and emotional processes of selectively paying attention and being available to concentrate on the communication received. The purpose of attending is twofold: (1) to prepare yourself to listen, and (2) to communicate to your clients that you are totally present and available to them. As a result of your attending, the client is encouraged to talk, the client is more likely to feel acceptance and positive regard, and you as the helper can listen and recall more easily.

When you attend mentally, certain physical behaviors generally follow. The behaviors listed below typically correspond to and communicate a message of mental attention:

- Facial expression reflecting interest; relaxed muscles

- Body position reflecting an open posture; slight lean toward the client; proximity

- Head nodding

- Eye contact reflecting appropriate respect and interest

❖ **EXERCISE 45: ATTENDING**

In a dyad, listen to a partner who is sharing a concern with you. Practice the attending behaviors listed above. Solicit feedback, and do a self-assessment of your ability to communicate to your partner your physical, mental, and emotional readiness for listening. Consider the following:

Mental and Emotional Readiness to Attend:

Nonverbal Behaviors (facial expression, body position, head nodding, eye contact, and other nonverbal behaviors):

CENTERING

The second step in the communication model is centering (see Table 6.1). Centering is the process of checking in with yourself in order to become congruent. It involves three steps: (1) listening to self, (2) being aware of your thoughts, feelings, and behaviors, and (3) differentiating between yourself and others.

Just as you are not in a position to deal with an aggressor physically if you are not balanced, you are not in a position to deal with and help a client if you are not first centered. Genuineness, positive regard, and boundary distinction are integrally connected to centering.

Genuineness

Genuineness means being your real, authentic self in the presence of others. It means being yourself rather than presenting a "mask" of what you think others want to see and will accept.

Congruence is a prerequisite to genuineness. Congruence means you *are* yourself. It means consistency of thoughts, feelings, and behaviors and an integration of your external image and internal experience. Congruence begins with self-awareness. It means allowing yourself to be who you are in the moment. If you can't be yourself with yourself, how can you be yourself with others?

❖ **EXERCISE 46: GENUINENESS**

Consider a recent interaction. To what extent were you being genuine? In what ways would you like to have been more genuine? What keeps you from being more genuine?

❖ **EXERCISE 47: BLOCKS TO GENUINENESS**

Potential blocks to genuineness include hiding behind a mask or professional role, unnaturalness, rigidity, lack of self-acceptance, or incongruence. Form a dyad with a partner. Do a self-assessment of blocks to your genuineness.

Positive Regard

Positive regard is the communication that you accept your clients for who they are, including their rights to their own feelings, opinions, and choices (short of harm to others). Giving positive regard does not mean you have to like, agree with, or condone behaviors. It is an acceptance of others' rights to be different from you.

❖ Exercise 48: Positive Regard

List 10 people you know (including a range of people you both like and dislike) and with whom you both agree and disagree. Include yourself on this list.

1.

2.

3.

4.

5.

6.

7.

8.

9.

10.

Now do a self-assessment of your ability to give positive regard to each of these people. Remember, positive regard means acceptance of individuals' rights to be themselves. It does not mean liking or agreeing with them.

What gets in your way of giving positive regard to those for whom you find it difficult? How would you be able to help individuals without giving them positive regard?

❖ **EXERCISE 49:   CHALLENGES TO POSITIVE REGARD**

Criticizing, imposing expectations, advice giving, telling the client what to do, and blaming interfere with communicating positive regard. Form a dyad. Have your partner share perspectives that you disagree with. Practice responding with positive regard. Assess your ability to give it without the above interferences.

Boundary Distinction

Boundary distinction is the ability to differentiate *your* thoughts, feelings, actions, issues, and problems from *other people's* thoughts, feelings, actions, issues, and problems. Boundary distinction is particularly challenging when you are invested in or attached to the image of another person. The woman who feels her partner is a reflection of her, for example, or the parents who see their child as an extension of themselves often find it difficult to distinguish where they end and the other person begins.

The purpose of the helping process is to focus on the needs of the client. The ability to distinguish boundaries, to recognize and separate your issues from your clients' issues, is an essential skill in being able to focus on the needs of the client. Boundary distinction is a natural outcome of the centering process. Focusing on distinguishing boundaries also facilitates the centering process.

❖ **EXERCISE 50:   BOUNDARY DISTINCTION**

Identify whose issues are whose in the excerpt below. Who owns what problems?

> Cindy recently moved in with Bill. They have talked at length about living abroad at some point but have not agreed on where or when.
>
> Cindy, who has been unemployed, was just offered a job that would involve a minimum of a two-year contract commitment, with a potential of it lasting for five years. She accepted the offer, committing herself to staying in their present location, without consulting Bill.

When Cindy shared the news with Bill, anticipating enthusiastic congratulations, she was stunned at Bill's angry response. He accused her of making a decision impacting both of them without getting input from him.

Cindy feels he's being unreasonable and controlling. In fact, she feels like Bill's trying to take control of her life, and is wondering whether their relationship is such a good match after all.

What's Bill's issue? (Bill's issue must relate to Bill.) What's the external event that disappointed Bill? Why is it a problem for him?

What's Cindy's issue? (Cindy's issue must relate to Cindy.) What's the external event that disappointed Cindy? Why is it a problem for her?

Give an example of a blurred boundary. (e.g., "If you really loved me you would . . ."):

Give an example of a clear boundary.

EMPATHIZING

The third and last step of the communication model in Stage 1 is empathizing (see Table 6.1). Empathizing means understanding the perspective of your clients and communicating that understanding to them. Empathizing is being able to look at the situation through your clients' eyes. Specific skills that help you empathize in Stage 1 include explicit empathy and concreteness.

Explicit Empathy

Explicit empathy is communicating an understanding of what the other person has stated directly. It can focus on thoughts, feelings, and/or behavior.

Paraphrasing is simply *rephrasing* what was said and does not necessarily include an understanding. It may be used as a tool to communicate empathy. Empathy, itself, however, means communicating an *understanding* of the other person's perspective.

Explicit empathy statements may include paraphrasing the other person's stated thoughts, feelings, and/or behaviors. For example:

> CLIENT: *I'm so nervous about this recital that I haven't been able to sleep for days. All I can think about is walking up in front of a filled auditorium and being scared to death. I just keep replaying it in my mind.*
>
> EXPLICIT EMPATHY (FOCUSED ON THOUGHTS): *So all you can think about is the recital, and particularly imagining walking to the front of the auditorium; and it just keeps replaying over and over again in your head....*
>
> EXPLICIT EMPATHY (FOCUSED ON FEELINGS): *It sounds as though you're feeling pretty anxious about this recital, both now, and anticipating being 'scared to death' when it happens....*
>
> EXPLICIT EMPATHY (FOCUSED ON BEHAVIOR): *So you haven't been able to eat or sleep for days....*

❖ EXERCISE 51:   EXPLICIT EMPATHY: THOUGHTS

Respond to the following excerpts using explicit empathy statements focused on thoughts:

1.   "What's the use? I can't seem to succeed at anything. My wife thinks I'm a failure. My father thinks I'm a failure. My kids don't even know me. I'm beginning to think they're right."

2.   "I am so discouraged. I guess I'm just not cut out to go into medicine. I really want to be able to help people and save lives, but I can't get through the labs. I practically faint at the sight of blood."

❖ EXERCISE 52:   EXPLICIT EMPATHY: FEELINGS

Respond to the following excerpts using explicit empathy statements focused on feelings:

1.   "I am so irritated. This is the second time this week that my supervisor has called me 'sweet cakes.' It's like he doesn't have a clue how disgusting that is."

2.   *"I've just about had it. If my boyfriend uses that tone of voice once more with
     me . . . I can't believe he doesn't know how much it hurts to have him talk to
     me like I'm a child."*

❖ **EXERCISE 53:   EXPLICIT EMPATHY: BEHAVIORS**

*Respond to the following excerpts using explicit empathy focused on behav-
iors:*

1.   *"I don't have a clue what to do about summer. I have three classes left to take
     to finish my degree, and I need to work to get some money. But I really feel
     like I need a break."*

2.   *"I can't believe I actually slapped Billy. He was whining and throwing one of
     his tantrums, which I can usually handle. I guess it was the last straw after
     a horrible day at work. I just snapped."*

Explicit empathy statements often appropriately need to respond to a combination of thoughts, feelings, and behaviors, that is, thoughts *and* feelings, feelings
and behaviors, or all three. An empathetic response to the client statement above
demonstrating such a combination might be this:

*"So you're really anxious and scared about this recital (feelings), to the point that you
just keep imagining it happening (thoughts), and can't even eat or sleep (behavior).*

❖ **Exercise 54:   Explicit Empathy: Responding to a Combination of Thoughts, Feelings, and/or Behaviors**

Respond to the following excerpts using explicit empathy statements focused on the thought, feeling, and/or behavior combinations reflected in each. Identify the focus by labeling as in the above example:

1. **Woman, age 22:**

   "I swear . . . if my brother brings up the topic of religion one more time, I'm going to scream. Why can't he just leave me alone and accept the fact that we hold different beliefs?"

2. **Woman, age 27, university student:**

   "He really bugs me. He behaves as though I'm his best buddy because we had one class together. Every time I see him, he acts like I'm a long, lost friend from years past. It's really driving me nuts. Besides that, he's always going out of his way to touch me— my arm, my shoulder, my back. It makes me extremely uncomfortable. If he touches me again, I'm going to turn around and sock him!"

3. **Man, age 25, outdoorsman:**

   "I'm so torn up inside. The vet told me that I should have my dog put to sleep. She's an older dog with problems, I admit, but she's been with me since I was a kid. I can't imagine putting her to sleep. That would be like murdering my best friend!"

4. WOMAN, AGE 37, MOTHER OF AN 11-YEAR-OLD BOY:

   "My son is in the fifth grade, and already I feel like he's growing away from me. I expected something like this to happen when he became a teenager, but he's still so young! I just don't know how to connect with him anymore. His clothes, his friends, his language—they're all so foreign to me. I just don't know what to do."

5. MAN, AGE 28, UNIVERSITY STUDENT:

   "If that woman in class interrupts me one more time, I'm going to walk out! I can't believe that someone could be so consistently rude! And the professor does absolutely nothing about it! If I were teaching the class, I'd give that woman a piece of my mind—I'd let her know how distracting she is, and how unappreciated her little anecdotes are!"

6. MAN, AGE 34:

   "I really miss my dad. We never got a chance to resolve the tension between us before he died. As far as he knew when he died, I hated his guts and never wanted to see him again. That's the last thing I said to him. I never got a chance to take it back and tell him how I really feel."

Concreteness

Concreteness is specifying and making clear a general, vague, or abstract feeling, thought, or behavior. Often, clarifying is half of resolving the problem. "I feel awful," for example, is vague and general. Stating "I feel like I'm incompetent because I failed my chemistry test" specifies the issue. It's hard to know what to do next until the issue is stated explicitly. This person might decide, for example, to drop chemistry, study harder, change majors, or seek tutoring, but he or she must decide what the issue is before making a decision on how to address it.

❖ **EXERCISE 55:** **CONCRETENESS**

Each of the three examples below presents a vague statement. Extracted on the next lines are the parts of the statement that need to be clarified before the problem can be identified and addressed. Using these examples as a guide, read the four numbered statements following, and select from them the words that need to be restated in specific terms so the problem can be clearly identified.

For example:

"My life's a mess and I don't know what to do about it."

    "My life's a mess"

"I did something really awful and now I'm sure I'm going to pay a high price for it. I'm scared to death."

    "something really awful"

    "a high price"

"I'm just a wreck. I feel totally out of it. I'm afraid I might lose it."

    "just a wreck"

    "totally out of it"

    "lose it"

1. WOMAN, AGE 42, HOUSEWIFE:

   "I feel so out of control. I'm eating everything in sight, whether I'm hungry or not. I just can't help myself. And now, my husband is starting to bug me about the weight I'm putting on, and he can turn into a real fiend."

2. WOMAN, AGE 37, EMPLOYEE OF A STATE AGENCY:

   "I can't believe how much I've overcommitted myself to work. But there's no way to back out. My supervisor is being a real jerk."

3. MAN, AGE 22, UNIVERSITY STUDENT:

   "I'm so disappointed in myself. I've never done so poorly on an exam in my life. This is going to destroy my G.P.A. I guess it goes to show that I really am a lousy student."

4. MAN, AGE 59, INSURANCE SALESMAN:

   "What is it with women nowadays? They want to have everything. My wife wants to have her cake and eat it, too! Well, not with me, I told her. She says she's bored—what a joke! It's not like there's not enough at home to keep her busy."

STAGE 1: RELATIONSHIP GOALS

In Stage 1 of the helping relationship, our goal for the relationship is the development of rapport (see Helping Relationship Goals in Table 6.1). Rapport means having a sense of harmony, internal calm, and accord. It indicates a balanced interrelationship.

❖ EXERCISE 56:   RAPPORT

Rapport can be facilitated by the (1) communication of the three R's, and (2) implementation of the Stage 1 facilitative skills. Form a dyad with a partner. Then solicit feedback and do a self-analysis of your effectiveness with each of the following. Give specific examples.

Communication of the Three R's

Projects an Attitude Supporting One's Right to Be Who One Is:

Projects an Attitude of Respect:

Projects an Attitude Supporting Appropriate Responsibility-Taking:

Stage 1 Facilitator Skills:

Listen:

Attending:

Center:

Genuineness:

Positive Regard:

Boundary Distinction:

Empathize:

Explicit Empathy:

Concreteness:

Stage 2 Helping Skills

In Stage 2 of the helping process, the goal is to help the client explore the underlying issues of the presenting problem. Exploration is promoted through examining the difference between the disappointments of external events and internal issues, and developing new perspective and insights. The goal of the helping relationship is processing and understanding.

Helper skills that can be used as a guide to examining underlying issues through processing and understanding include step 4 of the communication model and its corresponding facilitative skills:

Communication Model Step   *Facilitative Skill*

4. Focusing   Implicit Empathy
                                     Confrontation
                                     Self-Disclosure
                                     Immediacy

FOCUSING

Focusing is the intentional directing of attention to specific aspects of issues for the purpose of facilitating clarification, insight, and understanding (see Table 6.1). Focusing can be accomplished through empathic focusing (communicating an understanding of the perspective of the person you're talking to) or expressive focusing (communicating your own perspective). You can focus on the explicit and/or implicit message of any one, or combination, of three general areas: (1) human dimensions (thoughts, feelings, and/or behaviors), (2) verbal behavior, and (3) nonverbal behavior.

Empathic Focusing

Empathic focusing results from using empathy. It is directing attention to the client's thoughts, feelings, and/or behavior by communicating an understanding of the *client's* perspective. Two specific skills that use empathic focusing are implicit empathy and confrontation (see Helper Skills, Table 1).

Implicit Empathy

While explicit empathy, used in Stage 1, communicates understanding of what the client has *explicitly* stated, verbally or nonverbally, regarding thoughts, feelings, or behaviors, implicit empathy focuses on the *unspoken* message. Implicit empathy is communicating an understanding of what the client has only implied. Implicit empathy is looking through the clients' eyes and helping them clarify their own perspectives. For example, "I'm not sure I can make it to the end of the semester" may imply that the speaker may be considering dropping out, even though that wasn't stated directly. In other words, the clients themselves may be unclear, vague, or confused about their direction or desires.

The difference between explicit and implicit empathy is that explicit empathy communicates an understanding of what the other person has *directly* stated whereas implicit empathy communicates an understanding of what the client has only *indirectly* implied. In formulating implicit empathy responses, consider three steps:

1.   *Explicit empathy:* It's helpful to communicate an understanding of what your client *has* stated before going on with implicit empathy.
2.   *The implied message:* You need to identify the message that has been implied before trying to formulate your response.
3.   *Implicit empathy:* Make a response that incorporates the explicit empathy and the implied message and present it as a tentative understanding of your client's perspective. Your client will let you know how accurate you are.

Example:

A single mother, 52, is talking about her relationship with her 17-year-old daughter. She wants her daughter to be responsible—to come home by 10 P.M. on week nights and by 1 A.M. on weekends. She has not yet recognized that she is a role model for her daughter. She tells you: "I just don't see why it is such a big deal. Kids have got to have some basic rules and then be responsible for them. I'm only asking her to be in by 10 P.M. on school nights and 1 A.M. on weekends. That seems reasonable to me, and she's even agreed. So why can't she do it? When she gets older she can come in a little later, as I do, but until then, I don't think it's too much to expect to have her home earlier."

1. *Explicit empathy:* "So even though your daughter has agreed to be in by those times, she doesn't comply, and you feel it's reasonable to expect that she should be responsible for getting home by then."
2. *Implied message:* It's okay for the mother to come home later and still expect her daughter to get in earlier.
3. *Implicit empathy:* "So even though the two of you have agreed to the times your daughter should be home, she still isn't getting home on time; and it sounds like you think it's reasonable for the time you come home to be different from the time your daughter comes home."

Example:

A young woman, 28, is dissatisfied with her marriage. She's explaining how the relationship is not meeting her needs. "I think I probably married too young. I didn't know who I was or what I wanted in life. Now I'm finding myself in a marriage with a man with whom I share no interests or values. He's never home, we seldom spend time together, and I'm wondering if we even have a marriage."

1. *Explicit empathy:* "So after having married young you're now finding that you're in a marriage with someone with whom you share very little—including time."
2. *Implied message:* There is no longer a marriage; the marriage is over.
3. *Implicit empathy:* "It almost sounds as though, after recognizing how little you have in common any more, you've come to a realization that the marriage may be over...."

❖ **Exercise 57:   Implicit Empathy**

In each of the following examples, respond with (1) explicit empathy, (2) identify the implied message, and (3) implicit empathy:

1. Young man, age 28, trying to go to school and work part-time. He's just failed a class and has lost his job:

"What's the use? I can't seem to do anything right. Everybody seems to see me as a failure. I never thought of myself as a failure before . . . but, now, I don't know. My life feels so dead; sometimes I wonder if I may as well be."

Explicit empathy:

Implied message:

Implicit empathy:

2. A WOMAN, AGE 46, TALKING ABOUT AN INTERACTION WITH HER HUSBAND, DURING WHICH HE FELL ASLEEP:

"In the middle of my talking about something really important to me I looked over and his eyes were closed. He was just dozing off. I just said 'never mind, we can talk about it later,' because I knew he has been working really hard and was tired. But I felt hurt. I ended up feeling like I don't matter."

Explicit empathy:

Implied message:

Implicit empathy:

3. *A CAMPUS MINISTER, AGE 36, TALKING TO YOU ABOUT QUESTIONS SHE IS RAISING CONCERN-ING THE POSITION SHE HAS HAD FOR THE PAST ONE AND ONE-HALF YEARS:*

    "I just don't know what it is. This is what I always thought I wanted to do. I'm in a location I like. I'm at a university that's progressive and prestigious. Now that I've made it, you'd think I'd be content instead of thinking about the next challenge or goal. That just doesn't seem like the kind of thinking a minister should have."

Explicit empathy:

Implied message:

Implicit empathy:

4. *WOMAN, AGE 38, SINGLE MOTHER OF ONE CHILD, TALKING TO A COUNSELOR:*

    "I just don't understand it. We've worked so hard with Billy on his schoolwork, but he still has trouble. The teacher told us yesterday that she wants to have Billy tested for a learning disability. I can't believe that is necessary! We'll just have to work a little harder, that's all. There's nothing you can't overcome with a little hard work. Besides, my son's no dummy. He takes after me, not his father!"

Explicit empathy:

Implied message:

Implicit empathy:

5. MAN, AGE 39, TALKING WITH A FRIEND:

"My mother is about to make me crazy! Here I am, a grown man, yet she insists on calling me daily, coming to my house with prepared meals, and trying to fix me up with any 'sweet young lady' she meets on the street! I'm ready to move out of state without telling her, just to get some space."

Explicit empathy:

Implied message:

Implicit empathy:

6. MAN, AGE 66, TALKING WITH HIS WIFE OF 40 YEARS:

"I've been trying to learn how to use the computer, but I just can't seem to get the hang of it. What's even more frustrating is that everything is run by computers these days. You can barely go shopping without having to interact with some kind of a computer. There's just no way I can keep up. I don't see what's so great about computers, anyway. Look at what's happened to our world as a result of mass communication! I miss the good ol' days, and I think 'progress' might well mean going back to them! At least for one man!"

Explicit empathy:

Implied message:

Implicit empathy:

Potential Problem Areas for Implicit Empathy

There can be difficulties in expressing implicit empathy. The following examples illustrate some inappropriate responses to the client statement "I think I need to get out of this relationship."

1. *Interpreting* (imposing your perspective): "You seem to be experiencing acute depression."
2. *Inaccurate* empathy (misunderstanding): "So you're really worried about what others think of you."
3. *Judgmentalness* (imposing your negative judgment): "So you've ended up feeling pretty sorry for yourself, haven't you?"

❖ EXERCISE 58: IMPLICIT EMPATHY ASSESSMENT

Assess your responses in the preceding exercise for interpreting, inaccurate empathy, and judgmentalness. Rewrite your responses, if necessary.

Confrontation

Confrontation, in this model, is a form of empathic focusing and a specific type of implicit empathy. Contrary to the typical view as a clashing of ideas between two people, confrontation here involves the helper's *empathizing* to help clarify clients' perspectives.

Confrontation is empathy that focuses on and helps clarify contradictions or discrepancies in client statements. For example, your client says that she wants her 20-year-old son to "be his own person and make his own decisions," but later she also states that "he better not drop out of college." Using empathic focusing on these contradictions and discrepancies helps the client clarifiy her underlying issues. For example, "So, it sounds like there are two things that are quite important to you. You want your son to be both himself and responsible for his own decisions, *and* you want him to choose to stay in college." Confrontation, then, is empathically focusing on contradictions and discrepancies from the client's perspective.

In using confrontation you need to (1) identify the discrepancy and (2) make an *empathic* statement that invites exploration of the discrepancy.

❖ EXERCISE 59:   SELF-CONFRONTATION

Identify two discrepancies in your own life to which you can respond with confrontations.

Example:

Discrepancy: I want to learn Spanish but I keep dropping out of class, and I don't study consistently while I'm in class.

Confrontation: "So, on the one hand, I would like to learn Spanish, but on the other hand, I seem to have a hard time committing in a way that results in learning it. It seems there are always other things that are more important."

Example:

Discrepancy: I want to be in a serious, committed relationship, but I keep getting involved with people who say from the beginning they're not interested in a commitment.

Confrontation: "So while I think I want to be in a committed relationship, for some reason I keep finding myself drawn to people who want to stay independent. Apparently there's some reason for the attraction."

1. *Discrepancy:*

    Confrontation:

2. *Discrepancy:*

    Confrontation:

❖ EXERCISE 60:   CONFRONTATION

Identify the discrepancies and write a confrontation for the following excerpts:

Example:

WOMAN, AGE 32, TALKING ABOUT HER POLITICIAN HUSBAND:

"*You know, I am so proud of him. He's really fought hard to get where he is, and he's had a lot of disappointments. It takes a lot of hard work and sacrifices, sometimes including your family. It takes a willingness to go on regardless of what your family wants. I just don't know why I had to be one of the sacrifices.*"

Discrepancy: Husband needs to sacrifice many things, sometimes including his family, to succeed, but he shouldn't have to sacrifice me.

Confrontation: "So, while you're proud of your husband's success, and understand it involves sacrifices—even of family, you don't understand why you, despite being family, would have to be one of the sacrifices."

1. GIRL, AGE 13, TALKING ABOUT HER RELATIONSHIP WITH HER MOTHER:

   "I really love her, you know. She is so cool. She gets on TV to talk and is always doing stuff that people look up to her for. Sometimes she's so busy I don't get to see her much. Sometimes it seems like she doesn't have time for me, or like I'm not very important. Sometimes I just hate her."

Discrepancy:

Confrontation:

2. MAN, AGE 45, TALKING TO A FRIEND ABOUT HIS WIFE:

   "Did you hear what my wife just said? She was asked a simple question about her job, and she makes a sarcastic remark about her boss. It was embarrassing. Now, I've told her how much I want her to have a good time at these parties, which I know she doesn't particularly enjoy. I've told her to just relax and be herself and have fun. Why does she have to get so serious?"

Discrepancy:

Confrontation:

3. Woman, age 43, single mother of three children, and university student:

"I'm really not that busy—it's all in a day's work. My schedule is pretty well set, which is how I like it. After I make breakfast and send the kids off to school, I clean the house and get cleaned up myself. I usually spend most of the day at the university taking classes. After class, I pick up the kids from my mother's house, go home, prepare dinner, and eat. I have time to study after they're in bed. I just want to have more unscheduled, free time!"

Discrepancy:

Confrontation:

4. Woman, age 53, talking about her ailing mother, age 84:

"This just makes me feel horrible. I mean what kind of a person would consider putting her own mother in a nursing home? What is wrong with me that I can't take care of her? I should take care of her. After all, she has devoted her entire life to taking care of me and my brothers. But my brothers act like there's absolutely nothing wrong with it. They don't offer to help. I just can't do it anymore."

Discrepancy:

Confrontation:

5. Woman, age 29, in a relationship of three years:

"I don't know what to do. I love him, and want to stay together, but the problems in our relationship seem to be multiplying exponentially. They're taking up almost all our time together. If we're not arguing, we're having serious discussions about what to do about all the issues. To be quite frank, I'm about at my limit. This is not how it *used to be. What's happened? What we have now is definitely not what I want from a relationship."*

Discrepancy:

Confrontation:

6.   WOMAN, AGE 29, GRADUATE STUDENT IN A COUNSELING PROGRAM, TALKING TO A FELLOW STUDENT:

"I'm unsure what to do, if anything. We've been learning about dual relationships in my ethics class, and it occurred to me that one of my professors has all kinds of dual relationships with his practicum students. In fact, I heard a rumor that he was sleeping with one of my classmates. It seems like I should say something or do something. But if I say anything, it might affect my grade. I don't know what to do."

Discrepancy:

Confrontation:

Potential Problem Areas for Confrontation:

Following are examples of some potential challenges in using confrontation effectively.

1.   *Judgmentalness:* For example, to your client, who says, "I really want to learn Spanish, but it seems like other commitments keep getting in the way," you respond: "Look, if you *really* wanted to learn Spanish, you wouldn't let other commitments get in the way." This response imposes a judgment.

2. *Not acknowledging discrepancies:* "So you really want to learn Spanish? That sounds like a great idea." This response doesn't acknowledge the discrepancy.

3. *Failure to empathize:* "To learn a language—any language—you really have to make a commitment. You just can't do it by studying once a week." This response fails to empathize with the dilemma of conflicting desires the client is experiencing.

4. *Imposing conclusions or assumptions:* "Okay, if you really want to learn Spanish, let's figure out a plan that will help you make that commitment and stick to it." This response *assumes* that the client wants to prioritize learning Spanish over the "other commitments" that have been "getting in the way."

❖ **EXERCISE 61:   CONFRONTATION ASSESSMENT**

Assess your responses from the previous exercise for judgmentalness, lack of acknowledgment of discrepancies, failure to empathize, and imposing conclusions or assumptions. Rewrite your responses, if necessary.

Expressive Focusing

Whereas empathic focusing communicates an understanding of the *client's* perspective, expressive focusing directs attention to the client's thoughts, feelings, and behaviors by communicating an understanding of the *helper's* perspective. The intent, however, is still the same: to direct attention to the *client's* thoughts, feelings, and/or behavior. The purpose is to help the *client* gain self-understanding by clarifying the issues.

Two specific skills that use expressive focusing are self-disclosure and immediacy (see Helper Skills, Table 6.1).

Self-Disclosure

Self-disclosure, literally, is disclosing yourself to another. A certain amount of indirect self-disclosure automatically happens through interaction. In a helping relationship, helpers disclose themselves indirectly through their congruency, openness, and genuineness. In other words, by *being* yourself, you are indirectly disclosing characteristics about yourself. Direct self-disclosure is the purposeful sharing of information about yourself that the other person wouldn't otherwise know.

When it is used to facilitate focusing, effective self-disclosure is the intentional sharing of information about *yourself* to promote client self-understanding by directing attention to the *client's* thoughts, feelings, and/or behaviors. If self-disclosure results in focusing on the *helper*, it is ineffective.

Example:

- A client states: "My supervisor at work drives me nuts! It seems as though I can't do anything to please her. She's always nit-picking. The problem is, I like my job otherwise, and I like my co-workers. I'm just not sure it's worth having to deal with her to stay there."

- *Ineffective self-disclosure*: "I know just what you mean. In fact, I'm in a situation right now just like that. My supervisor not only nit-picks on me but also seems to enjoy giving me all the assignments nobody wants. Like the other day, she asked me to do an inventory of all the office equipment! That's not part of my job description!"

- *Effective self-disclosure*: "One time I was in a position where my supervisor and I seemed to have differing perspectives on almost everything. It seemed I had to decide whether I wanted to invest energy in trying to reconcile our differences, leave, or just accept how she dealt with the differences. I'm wondering if you're experiencing anything like that in your situation."

Effective self-disclosure results in role modeling and validation of feelings, and gives the client permission to explore potentially threatening feelings. A simple self-disclosure— "When I was in a similar situation, I felt really inadequate and scared, and I'm wondering if you feel anything like that?"—invites the client to acknowledge and explore similar feelings.

❖ **EXERCISE 62:   SELF-DISCLOSURE**

Using the examples of client statements from Exercise 60: Confrontation, choose three and respond to them with ineffective and effective self-disclosure.

Example:

Ineffective self-disclosure: "I know just what you mean. I was married to a U.S. senator for 10 years, and our marriage finally just couldn't take it. A marriage can't last if it is not a priority."

Effective self-disclosure: "I think I can relate. When I was married to a U.S. senator, I finally had to sit down and identify as honestly as I could what I really thought was most important between conflicting issues. I'm wondering if that's where you are?"

1.   *Ineffective self-disclosure:*

    Effective self-disclosure:

2.   *Ineffective self-disclosure:*

    Effective self-disclosure:

3.   *Ineffective self-disclosure:*

Effective self-disclosure:

Potential Problem Areas in Using Self-Disclosure:

Following are examples of some potential challenges in using self-disclosure effectively.

1. *Distracting or overwhelming self-disclosure*: In the case above of the client with the "nit-picking" supervisor, the helper's ineffective self-disclosure distracts the client from her situation.

2. *Irrelevant disclosure*: Using the "nit-picking" supervisor case, an example of irrelevant disclosure might be, "I, myself, hate having to go to any kind of work."

3. *Focus shifts to helper*: In the above case, a response from the client of "Oh, really? So if that wasn't part of your job description, what did you do?" would reflect a shift in focus to the helper.

❖ EXERCISE 63:   SELF-DISCLOSURE ASSESSMENT

Assess your responses in the previous exercise for distracting, irrelevant, or focus-shifting self-disclosure. Rewrite any ineffective responses, if necessary.

❖ EXERCISE 64: SELF-DISCLOSURE DYAD

Form a dyad with a partner. Practice using effective self-disclosure to focus on client issues. Solicit feedback and do a self-assessment of the effectiveness of your self-disclosure.

Immediacy

Immediacy is a skill used to focus on the dynamics within the immediate relationship (e.g., helper-client). Immediacy uses both helper perspective (expressive), and client perspective (empathic). The purpose of immediacy is to direct attention to the dynamics in the current relationship as a means of helping clients explore and better understand their identified issues. There are two types of immediacy: long-term dynamics and current dynamics. Both focus on the dynamics *within the relationship* as they relate to client issues.

Long-term dynamic immediacy focuses on the overall relationship.

Example:

- "One of the issues you've been concerned about has been how much you tend to defer to others, and you've said you want to stop doing that. I've been noticing that in our relationship I often feel like you defer to me in terms of what we focus on—homework assignments, and so forth. It seems that, ironically, even though you're here to try to stop deferring to others, it's happening in our relationship, too."

Current dynamic immediacy focuses on what's happening currently—right now—in the relationship.

Example:

- "Just now, when I asked where you'd like to start, you suggested we start wherever *I* thought would be a good follow-up to last session."

The purpose of using immediacy is to promote client self-understanding by examining underlying issues and gaining new perspective and insight. Specifically, immediacy invites the client to examine interpersonal dynamics within the context of a safe, accepting, and nonthreatening environment. Notice that, as with self-disclosure, it's important to relate the dynamic back to the context of the client issue.

❖ EXERCISE 65:   IMMEDIACY

In the following examples, based on the identified client issue, give potential examples of long-term dynamic immediacy, and current dynamic immediacy.

Example:

Issue: *Desire to please/fear of rejection*

Long-term Dynamic Immediacy:

> "I've noticed that over the last few months while I've been seeing you, you've been incredibly conscientious about not missing a session, being here on time, and always having your homework suggestions done. Since the issue we've been focusing on is your desire to please others in order to prevent their getting upset with you, I'm wondering if your conscientiousness in here is an example of what you've been talking about."

Issue: *Desire to please/fear of rejection*

Current Dynamic Immediacy:

> "You've talked a lot about how important it is for you to please others. Now, when I just talked about your conscientiousness in here and you said you thought I was right, I can't help but wonder if that's an example of your wanting to please. It appears to be a real dilemma: wanting to give yourself permission to stop having to please, but at the same time wanting the other person to be happy with you."

Example:

Issue: *A woman, 33, is talking about her relationship with her father. She feels her father still treats her like she's a 17-year-old child.*

Long-term Dynamic Immediacy:

> "You know, Dad, I know you've said it's been hard for you to see me as an adult—that I'm still your 'little girl.' I'm really feeling strongly, though, about wanting to have an 'adult-adult' relationship with you, and when things happen like your telling me to be back by a certain time, it feels like I'm being seen as a 17-year-old again, and then I respond like I am. I'm wondering if we can't come up with a way to communicate that works better for us."

Current Dynamic Immediacy:

> "Actually, what just happened is a pretty good example of what I want. I shared my concern with you, calmly and rationally—like an adult—and you responded with acknowledgment of the issue. That felt good to me."

1. **Client Issue:** *A man, 43, is talking to his counselor about his inability to let go of his dependency on others for approval. He has described in detail how he has always deferred to his mother, then to both of two successive wives. He is now asking you if you think he's doing the right thing by cutting off his relationship with his mother.*

Long-term Dynamic Immediacy:

Current Dynamic Immediacy:

2.   **Client Issue:** *You call in a 14-year-old girl who was referred to you by the school counselor. She's been in your office several times before for skipping classes and not handing in assignments. She has run away twice from home and has made it clear that no one is going to tell her what to do.*

Long-term Dynamic Immediacy:

Current Dynamic Immediacy:

❖ **EXERCISE 66:   IMMEDIACY APPLIED TO YOURSELF**

Think of a circumstance about which you have some unresolved issues with someone (a relative, friend, or family member). Identify the issue, then write a long-term dynamic and a current dynamic immediacy response as though you were talking to that person.

1.   *Circumstance and Issue:*

        Long-term Dynamic Immediacy:

        Current Dynamic Immediacy:

2.   *Circumstance and Issue:*

        Long-term Dynamic Immediacy:

Current Dynamic Immediacy:

Immediacy incorporates several other skills. It may contain elements of self-disclosure, empathy, and/or confrontation. For instance, looking at the first example given of long-term dynamic immediacy, you might identify the following segments as an example of each skill:

- *Empathy:* "One of the issues you've been concerned about has been how much you tend to defer to others and you've said you want to stop doing that."

- *Self-disclosure:* "I've been noticing that in our relationship I often feel like you defer to me in terms of what we focus on, homework assignments, and so forth."

- *Confrontation:* "It ironically seems that, even though you're here to try to stop deferring to others, it's happening in our relationship, too."

❖ **EXERCISE 67: IMMEDIACY ELEMENTS**

Identify elements of self-disclosure, empathy, and/or confrontation in one or two of the examples you wrote above.

1. *Elements of empathy:*

Elements of self-disclosure:

Elements of confrontation:

2.   *Elements of empathy:*

Elements of self-disclosure:

Elements of confrontation:

Potential Problem Areas in Using Immediacy

Following are some potential challenges in using immediacy effectively:

1.   *Ignoring Relationship Dynamics*: Simply not addressing the issue, or ignoring a parallel between what's happening in your relationship and the issue your client is talking about.
2.   *Judgmentalness*: Responding with a judgmental or sarcastic tone; for example, "You say you want to stop deferring to others, but here you are deferring to me."
3.   *Interpreting or Drawing Conclusions*: Making an assumption, or conclusion, for example, "You know, if you really want to stop deferring to others, the place to start is right here and now."

❖ **EXERCISE 68:   IMMEDIACY ASSESSMENT**

Assess your responses to the previous exercise. Check for ignoring dynamics, judgmentalness, and/or drawing conclusions. Rewrite responses, if necessary:

Examples of ignoring dynamics?

Examples of judgmentalness?

Examples of drawing conclusions?

❖ **Exercise 69:   Immediacy Dyad**

Set up a dyad in which you use immediacy. Solicit feedback and do a self-assessment of the effectiveness of your use of immediacy.

STAGE 2:   RELATIONSHIP GOALS

In the second stage of dealing with underlying issues, our goal for the relationship is an activity of processing, with the resulting outcome of understanding (see Relationship Goals, Table 6.1). Processing is an exploration of perspectives with the intent of facilitating understanding.

❖ **Exercise 70:   Processing**

Effective processing is facilitated by (1) a continued foundation of the 3 R's (rights, respect, and responsibility), and (2) the effective implementation of the Stage 2 facilitator skills. Set up a dyad with a partner. Then solicit feedback and do a self-assessment of your effectiveness with each of the following skills. Give specific examples.

Attitude Assessment:

Projects an Attitude Supportive of Rights

Projects an Attitude of Respect

Projects an Attitude Supporting Appropriate Responsibility-Taking

Assessment of Stage 2 Facilitator Skills

Focusing (*Empathic*):

Implicit Empathy

Confrontation

Focusing (*Expressive*):

Self-Disclosure

Immediacy

CHAPTER 18

Stage 3 Helping Skills

In Stage 3 of the helping process, the goal is to help the client establish direction and change. With new perspectives and insight developed in Stage 2, direction and change is promoted by helping the client determine direction and implement desired change (see Table 6.1). The goal of the helping relationship in Stage 3 is to provide support during this change, and ultimately to terminate the helping relationship.

Helping skills that can be used as a guide to facilitating direction and change include step 5 of the communication model, directional support, that was addressed in Part IV, and the corresponding facilitative skills that are addressed in this chapter:

Communication Model Step	*Facilitative Skills*
5. Directional Support	Directionality
	Implementational Support

PROVIDING DIRECTIONAL SUPPORT

Giving directional support is providing follow-through support as your client works on implementing direction and change. Two specific skills that can be used in providing directional support are directionality and implementational support.

Directionality

Directionality is the facilitation of movement in a purposeful direction. Although directionality actually begins to develop in Stages 1 and 2, a clearer sense of directionality should have emerged by Stage 3. If you feel as though you're just "spinning your wheels" and not getting anywhere, then maybe you're not. Perhaps you're lacking directionality.

Directionality emerges in large part as a result of effective helping and the effective use of Stage 1 and Stage 2 facilitative skills. By Stage 3, however, a sense of direction should be more clearly apparent.

WHY WE'RE MAKING SUPERB PROGRESS, WHY DO YOU ASK?

Consider the following two examples, the first reflecting a lack of directionality and the second reflecting directionality:

Example (lacks directionality):

CLIENT: *I can't stand my job anymore. I'm beginning to consider quitting. And on top of that, my girlfriend's breaking up with me, too. Life's the pits these days.*

COUNSELOR: *So your girlfriend's breaking up with you . . .*

CLIENT: *Yeah, it's been coming for a while. My mom's really more upset about it than I am. She thought my girlfriend walked on water.*

COUNSELOR: *It sounds like the breakup's been hard on your mother.*

CLIENT: *Yeah—well, actually, I guess my sister was pretty fond of her, too.*

Example (with directionality):

CLIENT: *I can't stand my job anymore. I'm beginning to consider quitting. And on top of that my girlfriend's breaking up with me, too. Life's the pits these days.*

COUNSELOR: *It sounds like you're feeling kind of down as a result of two potential big losses in your life at once: your job and your girlfriend.*

CLIENT: *Yeah. And it really does feel like losses, too—big ones. My girlfriend and I have been together for four years and I've had this job for three.*

COUNSELOR: *So they're two losses in which you had invested a lot of time and, it sounds like, energy.*

CLIENT: *Yeah. I don't know which feels worse. Both feel like cutting off an arm or something. I hate losing stuff, anyway. Ever since my Dad left us when I was a kid, I've been really sensitive to loss.*

Themes. One way to facilitate directionality is to look for underlying themes and respond to them. For instance, in the example above there was a theme of loss:

With further exploration, the theme appears to have started in the client's childhood with the loss of his dad.

❖ **EXERCISE 71:   DIRECTIONALITY—THEMES**

In the following excerpts, identify a (1) theme and (2) response using the theme or directionality.

1. WOMAN, AGE 45, TALKING ABOUT HER RELATIONSHIPS:

    "Sometimes I feel so hurt. It seems as though if I didn't call home, I'd never hear from my parents. They never call me. I've always been the one to keep in contact with them. I don't think I'd have a relationship with them if I didn't. And now it feels the same way with my boyfriend. If I didn't call, I'm not sure how much I'd see him."

Theme:

Directionality Response:

2. MAN, AGE 27, TALKING ABOUT HIS FATHER'S INFLUENCE:

    "I really think Dad means well. But sometimes he just drives me nuts. He insists that I should be an attorney. It doesn't matter whether I want to or not. He thinks he knows best. It's been that way ever since I can remember. Even when I was really little he used to sign me up for stuff—soccer and Little League—that I wasn't interested in. And he's still trying to run my life."

Theme:

Directionality Response:

Underlying Internal Issues: Directionality emerges as one moves from the external situation to underlying internal issues. In this case, it might look like the following:

- External: dad left   quit job   girlfriend breaks up
- Theme: loss
- Internal Issue: fear of abandonment

Example:

COUNSELOR: *So it sounds like your dad's leaving had a strong effect on you and that you've been more sensitive to loss—like losing your job and now your girlfriend—ever since. It seems as though loss is a particularly painful experience for you.*

CLIENT: *Yeah, it is. It seems like I'm so aware of it that I'm almost expecting it—and then I get even more anxious about it. I'm always worrying about whether I'm going to be left again. It happens almost as soon as I meet someone.*

❖ EXERCISE 72:   DIRECTIONALITY—INTERNAL ISSUES

Using the two examples from the above "Theme" exercise, identify a (1) potential internal underlying issue and (2) a directionality response.

1.   **Internal Issue:**

Directionality Response:

2. *Internal Issue:*

Directionality Response:

Implementational Support

Implementational support is assisting the client in developing and implementing a strategy for bringing about the identified desired change. General application strategy guidelines for providing implementational support include these:

1. Keep It Simple
2. Take Small Steps
3. Make It Replicable
4. Focus on Positive Action
5. Be Independent
6. Make Sure It's Reasonable

❖ **EXERCISE 73:   CHANGE STRATEGY**

Identify a desired change you would like to make. Apply the six guidelines in developing a strategy to implement the change:

Desired change:

1. *Keep It Simple:*

2. *Take Small Steps:*

3. *Make It Replicable:*

4. *Focus on Positive Action:*

5. *Be Independent:*

6. *Make Sure It's Reasonable:*

STAGE 3: RELATIONSHIP GOALS

In the third and final stage of direction and change, the goal for the relationship is directionality, with a resulting outcome of change (see Relationship Goals, Table 6.1). Directionality, as discussed above, is the emergence of a sense of purposeful direction. Directionality emerges not just in the processing of the client's issues but within the relationship itself. A sense of movement and purpose will appear within the context of the helper-client relationship, and there will be a resulting change within the relationship. Change will occur as the client moves toward more and more independence from the helper. Ultimately, the change will result in termination of the relationship.

CHAPTER 19

Integration and Application of Skills

The stage structure for the helping process (see Table 6.1) provides an organized approach for facilitating client self-understanding and growth toward a positive self-concept. The steps of the communication model and the facilitative skills provide guidelines for the helper.

These are only guidelines. The helper needs to use judgment in determining which communication and facilitative skills to use at given times and circumstances, and with which clients.

❖ **EXERCISE 74:   INTEGRATION OF SKILLS**

Form a dyad with a partner, spending a full 45–50 minutes responding to effectively facilitate client growth and self-understanding. Use as many helping skills as you can. Then solicit feedback and do a self-assessment of your effectiveness with attitudes, communication model steps, and facilitative skills in each of the three stages of the helping process. Use the Self-Evaluation Form that follows. Rate your use of each skill on a scale of 1 (ineffective use) to 5 (very effective use).

Framework for the Helping Process
SELF-EVALUATION FORM

Pre-Stage Attitudes

	Ineffective			Effective	
	1	2	3	4	5

Rights
Communicates an attitude that clients have a right to be who they are, and a right to their own beliefs, feelings, and choices.

Potential Problem Areas: *Specific Examples:*

Control:
Engaging in controlling behavior; overly invested in a specific outcome; trying to manipulate client.

Respect: 1   2   3   4   5

Communicates an attitude of respect for capability and worth of the client. Honors cultural, ethnic, racial, familial, gender-role, and individual differences.

Potential Problems Areas: *Specific Examples:*

Judgmentalness:
Imposing helper values and beliefs on the client; lack of distinction between client as a person and client behavior; nonverbal criticism.

Appropriate Responsibility: 1   2   3   4   5

Communicates an attitude and understanding that clients make their own decisions regarding who they are and how they live their lives, and are responsible for their own actions, choices, and change.

Potential Problem Areas: *Specific Examples:*

Rescuing:
Taking responsibility for the client's decisions; attempting to problem-solve or "fix" the client; taking responsibility for the client's behavior within the helping process.

Blaming:
Blaming the client for not moving fast enough; blaming the client for his or her circumstances or behavior.

Stage 1 Skills

*STEP 1.   LISTENING   1   2   3   4   5

Is able to listen effectively.

Potential Problem Areas: *Specific Examples*

Nonlistening, pretend listening; ineffective
selective listening; inappropriate self-
focused listening.

Attending   1   2   3   4   5

*Communicates to client a physical and
mental presence and readiness to concentrate on the client and client's needs.*

Potential Problem Areas: *Specific Examples*

Inattendance, as reflected by facial expression, body position, head movement, eye
contact, or other nonverbal behaviors or
verbal statements.

---

*STEP 2.   CENTERING   1   2   3   4   5

*Has self-awareness, congruence, and
internal emotional equilibrium, and ability
to differentiate own issues from client's.*

Potential Problem Areas: *Specific Examples*

Lack of self-awareness; incongruence;
emotional imbalance; defensiveness;
mental confusion; "button pushing."

Genuineness   1   2   3   4   5

*Being real, authentic, and congruent in the
presence of the client.*

Potential Problem Areas: *Specific Examples*

Hiding behind a mask or professional
"role"; unnaturalness; rigidity; lack of self-
acceptance; incongruence.

Positive Regard   1   2   3   4   5

*Communicates acceptance of the client as a
worthy human being. Acknowledges the
right of others to be different from you in
thoughts, feelings, and choices.*

*Communication Model Step

Potential Problem Areas:

Verbal criticism; imposing expectations of thoughts, feelings, and choices; advice giving; telling clients what they "should" do; blaming.

Specific Examples

Boundary Distinction   1   2   3   4   5

Makes effective distinction between own thoughts, feelings, behaviors, experience, and issues and those of client.

Potential Problem Areas:

Projection; defensiveness; inappropriate responsibility-taking; rescuing; blaming; controlling; judgmentalness.

Specific Examples

---

*STEP 3.   EMPATHIZING   1   2   3   4   5

Able to communicate an accurate understanding of the client's feelings, thoughts, and/or behaviors from the client's perspective.

Potential Problem Areas:

Inaccuracy; imposition of helper's perspective.

Specific Examples

Explicit Empathy   1   2   3   4   5

Communicates an accurate understanding of the client's thoughts, feelings, and/or behaviors from the client's perspective, based upon what the client actually states explicitly.

Potential Problem Areas:

Inaccuracy; imposing helper perspective; interpreting; allowing the client to ramble; inappropriate or inconsistent terminology; rambling.

Specific Examples

Concreteness   1   2   3   4   5

Helps client clarify and translate general and vague thoughts, feelings, and behaviors into specific, concrete examples.

Potential Problem Areas:

Allowing vague generalities; allowing or encouraging abstractions.

Specific Examples

*Communication Model Step

Stage 2 Skills

*STEP 4.   FOCUSING   *1   2   3   4   5*

Helps client identify and clarify issues.
Moves from initial problem identification
to problem differentiation, examination of
issues, and new perspectives and insights.

Potential Problem Areas: *Specific Examples*

Lack of clear problem differentiation; lack
of issue examination; lack of insight.

Implicit Empathy   *1   2   3   4   5*

Communicates an accurate understanding
of what client only implies, or has vaguely
formulated.

Potential Problem Areas: *Specific Examples*

Imposing helper perspective; inaccuracy;
judgmentalness; drawing conclusions;
problem solving for the client.

Confrontation   *1   2   3   4   5*

Communicates an accurate understanding
of client's experience of discrepancies in
thoughts, feelings, behaviors, and/or experi-
ence.

Potential Problem Areas: *Specific Examples*

Judgmentalness; ignoring discrepancies;
failure to empathize; imposing conclusions
or assumptions; blaming.

Self-Disclosure   *1   2   3   4   5*

Helper sharing of information about self in
a way that promotes client self-understand-
ing. Disclosure is effective and does not
distract or take the focus off the client.

Potential Problem Areas: *Specific Examples*

Distracting or overwhelming disclosure;
irrelevant disclosure; role reversal that
shifts the focus off the client and on to the
helper.

*Communication Model Step

Immediacy   *1*   *2*   *3*   *4*   *5*

Acknowledges and addresses dynamics within the helping relationship (either current or long term) in a way that helps facilitate exploration of client issues and self-understanding.

Potential Problem Areas: *Specific Examples*

Ignoring dynamics within the helping relationship; judgmentalness; interpreting or drawing conclusions; blaming; defensiveness.

Stage 3 Skills

* STEP 5. PROVIDING DIRECTIONAL SUPPORT   *1*   *2*   *3*   *4*   *5*

Helps client move from new perspectives and insight to implementing change.

Potential Problem Areas: *Specific Examples*

Lack of clear perspective and insight; lack of movement toward change.

Directionality   *1*   *2*   *3*   *4*   *5*

Helps client gain clarity of direction through problem differentiation, recognition of recurrent themes, and/or new insights.

Potential Problem Areas: *Specific Examples*

Lack of direction; wheel spinning; imposition of irrelevant helper perspective.

Implementational Support   *1*   *2*   *3*   *4*   *5*

Helps client put insight into action by implementing identified desired changes. Uses strategy identification, implementation, and assessment.

Potential Problem Areas: *Specific Examples*

Lack of direction; lack of strategy identification; lack of action.

*Communication Model Steps

*S*elf-Development Section

A Framework for
Personal Growth

How can you be a guide if you're not on the journey?

—ANONYMOUS

PART

I

Personal Growth
and a Positive Self-Concept:
Goals and Dilemmas

Personal growth is a process that implies movement toward an
outcome. That outcome goal, in this manual, is identified as
psychological health, self-fulfillment, and happiness as
evidenced by a positive self-concept.
A positive self-concept implies *internal* self-approval.
Many of us are conditioned to focus on *external* approval,
and we develop external selves or "masks" in an attempt to gain
acceptance. We subsequently experience a discrepancy between
our external "masks" and our internal experience of ourselves.

I remember admiring a girl named Allison in second grade. For a second grader she seemed extraordinarily confident of herself and unfaltering in her decisions about what to do. We both attended a two-room country school, and one day in early spring, she and I came upon a group of other children congregated behind the swing shed at school. In turn, they were each pulling down their pants or pulling up their skirts and twirling around for the others to see. It appeared to be a simple lesson in anatomy. I'll always remember how, while I was surprised, intrigued, and considering whether to leave, watch, or join this group, Allison went, without hesitation, directly to the teacher. The incident became a major community affair, with parents being notified, asked to come to school, and the identified kids being sent home in disgrace. I remember thinking that, through no credit of my own, I had accidentally been spared the humiliation levied at the other children. This was one of my first conscious realizations that life wasn't going to be as easy as I'd thought. There were going to be some pretty tough decisions when the right thing to do was apparently not going to be that obvious. I also became acutely aware that I had choices, that choices had consequences, and that I needed to make choices based on the outcome I wanted.

Personal Goals and
the Pursuit of Happiness

Many people struggle, saying they don't know how to get what they want. More likely, they have trouble *identifying* what they want, and subsequently, have trouble attaining it.

I have a vivid recollection of my college graduation party. I remember the image I had as I planned the party, of relaxing with friends and enjoying some time of frolic and fun. On the evening of the party, however, I became preoccupied with "hosting." I worried about whether everyone was having a good time and felt I couldn't spend too much time with any one guest. Generally, I felt stressed and was glad when it was over. I realized afterward that I had lost sight of my goal—what I had said I wanted. It was unconsciously, replaced with another goal. The goal I unconsciously substituted was that everyone else have a good time, and that I be perceived as a good hostess. These are valid goals, but they were not what I initially intended. My original goal was lost. I did not relax and enjoy the time with my friends.

Self-development and personal growth imply change and movement toward an outcome. Unless we are clear about that outcome, we will find that reaching it is difficult. Unless we understand what we want out of life, we will not easily achieve it.

❖ **EXERCISE 1:   PERSONAL GOALS**

What do you want out of life? What would you identify as your personal goals?

> *Happiness achieved a rank of fifth in a list of eighteen values or goals that were appraised for importance by a national cross-section of adult Americans.... In a similar appraisal with Midwestern college students as subjects, happiness was found to rank second, and students from three community colleges in California ranked happiness as first.   (ARKOFF, 1988, P. 259).*

Many people, when asked what they want out of life, say "happiness," or some variation, such as "self-fulfillment" or "life satisfaction." When asked, however, what they mean by "happiness," they have to stop and reflect.

❖ **EXERCISE 2:   HAPPINESS DEFINED**

Describe below what happiness means to you.

Often people define happiness as feeling good, or as achieving life satisfaction or self-fulfillment. It has been defined as "a state of well-being, and contentment; notably well adapted or fitting; a state of pleasurable contentment of the mind" (*Webster's New Collegiate Dictionary*, 1981). But how do you attain happiness, life-satisfaction, self-fulfillment, or a pleasurable contentment of the mind?

❖ EXERCISE 3:   ATTAINING HAPPINESS, LIFE SATISFACTION, AND
SELF-FULFILLMENT

List ten things that would bring you happiness, life satisfaction, and self-fulfillment.

1.

2.

3.

4.

5.

6.

7.

8.

9.

10.

People sometimes equate happiness with the attainment of something external: "I'd be happy if..."

- "...my house were paid off."
- "...I had a boss who wasn't such a dictator."
- "...my husband were more patient."
- "...my kids were more obedient."
- "...I had my degree."

- "...I could just have more time to myself."

- "...I had a new car...bigger house...new boss...better job."

How many of the ten things you identified above are external?

In seeking happiness, many people initially focus on the external world. They want a new car, a new job, an orderly house, a partner who cares and acts the way he or she "should," children who are bright and responsible—the list is endless. People do not always behave in the way we want them to, however, and events seldom happen as we would like. Consequently, when our happiness is dependent on people and things in the external world, we are often disappointed and unhappy.

❖ EXERCISE 4: DISAPPOINTMENT AND UNHAPPINESS

List some of the things that have made you unhappy recently. Then look at your list and put asterisks by those for which your unhappiness seemed to be the result of disappointment in people or things. How much of your unhappiness/happiness appears to be dependent on your external world?

The culture and environment from which we come and in which we live may reinforce ideas and ways of being that may result in the continuation of unhappiness, as discussed in Chapter 3. However, let's consider another perspective. Let's consider that happiness is *not* based on others, outside events, or what happens in the external world, but that happiness resides within.

> *Happiness is feeling satisfied with who you are and how you're living your life.*

A highlight of my childhood was the summer before I entered fourth grade. My family moved out to a little farm where neither my brother, who was going into sixth grade, nor I, knew anyone. For me that meant a wonderfully quiet summer with a hayloft, country roads, treehouse, stilts, a creek down across the field with moldable clay, and having my brother all to myself. The experience of summer was one of flowing through days, deciding along the way what to do, and letting the days unfold. There were no expectations that anything would "result" from a day but rather a sense that the day was there to be enjoyed, and I had a rich awareness of pleasure and contentment. It seemed clear that happiness meant enjoying the time. I became conscious during this time that happiness clearly must have something to do with feeling good about yourself and how you're spending your time. It had to do with my feeling about my experience.

CHAPTER 2

A Positive Self-Concept:
Internal Judgment and Approval

The goal of self-development and personal growth can be described with a variety of terms: *self-fulfillment, life-satisfaction, a positive self-concept, optimal development, self-realization, effective functioning, psychological health*, or as many might say, *happiness*. For continuity of terminology, the term *positive self-concept* is used in this manual to refer to the goal of personal growth.

Happiness was defined in the previous chapter as feeling satisfied with yourself and how you are living your life. In this chapter, we propose that the elements of this definition represent the foundation of a positive self-concept.

A POSITIVE SELF-CONCEPT

A positive self-concept, simply put, means you have a positive conceptualization or view of yourself. Following are three elements that provide the basis for a positive self-concept:

1. **Self-Perspective**

    This is how you see the world, based on *your* judgment, not somebody else's. The words *self-concept* embody the idea of the perspective of the *self*. They imply *internal* judgment—judgment and approval that can come only from within.

    While the perspectives of other people have an influence on our self-concept formation—particularly in the initial stages of our development—they are not the same as our self-concept.

    Self-concept is one's picture of himself or herself and one's own independent evaluation of this picture. (ADAPTED FROM BRAMMER, 1988).

2. **Conceptualization of the Self**

    Your self-concept is how you view yourself in the three basic conscious human dimensions: feelings, thoughts, and behaviors.

139

3. **Positivity**

A positive self-concept means experiencing yourself positively—as someone who is worthy, good, and acceptable.

A positive self-concept means that your judgment and conceptualization of yourself—including your feelings, thoughts, and behavior—are positive. A positive self-concept, then, is feeling satisfied with yourself and how you're living your life.

The most significant aspect of this definition is that it reflects the *individual's* judgment. A positive self-concept is based on *internal* approval. *You* are judging *yourself* positively.

❖ **EXERCISE 5: SELF-CONCEPT**

Describe how you feel about yourself. In what ways do you feel satisfied or dissatisfied with who you are?

Describe your feelings about how you're living your life. In what ways do you feel satisfied or dissatisfied? Do you enjoy how you spend your time?

I experienced a significant transition when I started ninth grade. For the first time in my academic career, grades would be used to determine my GPA, to decide whether I would get into college, and to become part of my permanent life's record. Suddenly, going to school, which I had always enjoyed, had to be taken "seriously." My accomplishments, after all, were to determine my very worth. Worse, others were going to be the judge. I became anxious, competitive, and obsessed with proving my worth.

CHAPTER 3

Conditioning toward External Judgment and Approval

Although a positive self-concept is based on *internal* judgment and approval, many of us are initially conditioned toward *external* judgment and approval. We are taught to focus externally—on people other than ourselves—for judgment and approval.

As children we are literally dependent on the acceptance of others—parents, teachers, religious leaders—for our survival and well-being. Consequently, we are experienced in focusing externally. Three major factors contribute to our focus on external judgment and approval:

1. The initial influence of external feedback on our self-concept formation
2. Our basic need of love and acceptance
3. Gender-role expectations

IMPORTANCE OF EXTERNAL FEEDBACK ON SELF-CONCEPT FORMATION

The image we develop of ourselves begins to form very early in our lives. Since others' responses to us initially constitute our primary experience of who we are, our conceptualization of ourselves is heavily dependent on how others react to us.

> *In our interaction with significant others we come, to a large extent, to see ourselves as [they] see us. We learn from others' reactions, our worth. We learn how competent, how attractive, how acceptable, how lovable we are.* (ARKOFF, 1988, P. 3)

> *In the socialization process the child is "filled up" with essence, as it were.* (LØVLIE, 1982, P. 73)

> *An infant coming into the world has no past, no experience in handling itself, no scale on which to judge his or her own worth. The baby must rely on experiences with people and their messages about his or her worth as a person. For the first five or six years, the child's self-concept is formed almost exclusively by the family.* (SATIR, 1988, P. 25)

Self-Concept Formulation: Positive or Negative.

Depending on the messages we hear, we can learn to conceptualize ourselves as positive, worthy, adequate, and lovable, or as negative, unworthy, inadequate, and unlovable. If we hear, "What a lovely, bright child you are; you are so clever," we learn to view ourselves as "lovely, bright, and clever." If, on the other hand, we hear, "You clumsy brat. This is the third time today you spilled your milk. Can't you do anything right?" We learn to view ourselves as "clumsy brats"—inept, and an inconvenience. Whether the message is positive or negative, we internalize it and it affects our view of ourselves. We often have difficulty sorting out what is truth about ourselves and what is simply inaccurate messages we have received.

❖ **EXERCISE 6:   MESSAGES**

What messages did you receive as a child that you perceived as positive?

What messages did you receive as a child that you perceived as negative?

Power of Initial Self-Concept

The self-concept, once formed, is your *experience* of *yourself* in the world. Your self-perceptions influence your feelings, thoughts, and behaviors. The self-concept holds over time and tends to find ways to reconfirm itself. This is true, regardless of whether your initial self-concept was positive or negative.

> *Outside forces tend to reinforce the feelings of worth or worthlessness the child learned at home: the confident youngster can weather many failures, in school or among peers; the child of low self-regard can experience many successes yet feel a gnawing doubt about his or her own value.*   (SATIR, 1988, P. 25)

> *Because of its influence on our perceptions, our self-concept influences our behavior in many ways. Some psychologists believe that the basic human force is striving to maintain or enhance our conception of ourselves.*   (ARKOFF, 1988, P. 4)

> *Our self-perceptions determine our behavior and attitudes. If we think we're small, limited, inadequate creatures, we behave that way. If we think we are adequate, healthy, and deserving, we behave that way.*   (WILLIAMSON, 1992, P. 59)

❖ EXERCISE 7:   SELF-CONCEPT FORMATION

What messages did you get as a child that still affect how you view yourself today?

How many of these messages are actually true?

At some point, we have to begin to look objectively at the early messages we received. From our own perspectives, we have to decide whether the messages we were given are accurate. We make decisions about what parts of the messages we will accept, what parts we will reject, and what parts of ourselves we may wish to change based on the messages we received. The first step is to assess how much of our self-concept is based on messages we received from others, and how much is based on how we actually experience ourselves.

❖ EXERCISE 8:   SELF-DESCRIPTION

Describe yourself based on how you think others perceive you.

Describe yourself based on how you perceive yourself.

Describe the difference between how you think others perceive you and how you perceive yourself.

OUR BASIC NEED OF LOVE AND ACCEPTANCE

According to Abraham Maslow (1968), the need for love and acceptance is part of a hierarchy of basic needs, and is more primary than either self-esteem or self-actualization. In other words, attaining love and acceptance initially takes priority over esteeming or actualizing yourself.

In attempting to gain love and acceptance, we initially focus externally on what we perceive that others want; we then try to accommodate these perceived desires. This stage is followed by continued external focus to assess others' reaction. In this process, we may sacrifice our internal perspective and judgment. For example, when a child hears "I'll have no back talk from you, young lady," when she is attempting to explain to her father that she didn't do what he is accusing her of, she may well stay quiet. In an effort to gain her father's approval, she may sacrifice sharing her perspective.

In our desire for acceptance and approval, many of us focus externally on what others think and want. We then react—either by behaving in ways to gain their acceptance, or by rebelling, for fear we wouldn't get their acceptance anyway.

❖ EXERCISE 9:   THE MOTIVATION OF ACCEPTANCE AND APPROVAL

Can you identify times from your childhood when you acted in a way to get someone else's acceptance and approval?

Can you think of times when you did something you disapproved of or disagreed with in order to get someone else's approval?

Can you identify times when you rebelled and did the opposite of what was expected of you?

How did you feel about yourself in each of these examples?

Being versus Doing

Our being is our innate essence—who we are. Our behavior is what we do. What we do can be changed. Our essence cannot.

A child hitting another child on the playground is exhibiting a behavior. The being is the child. A response, such as "You bad boy," is a judgment of the child's being. A response, such as "Hitting is unacceptable behavior," focuses on the behavior without stating a judgment about the being. We can disagree with and change behavior while still accepting the being.

❖ **EXERCISE 10:   BEING AND BEHAVIOR**

Can you identify times when you were judged on your behavior rather than your being?

Can you identify times when you judged yourself based on your behavior rather than your being?

Unconditional versus Conditional Acceptance

Unconditional acceptance means a suspension of judgment and a total acceptance of the individual's *being* without conditions or contingencies. It differentiates between who you are and what you do. You can disagree with an individual's behavior and still give unconditional acceptance to the individual's being.

Conditional acceptance means that acceptance of the individual is contingent upon his or her behavior. It means that acceptance is dependent on what the person does.

- "I'll love you *if* you don't spill your milk."

- "I'll love you *if* you're good and clean your room."

- "I'll be proud of you *if* you get straight A's."

- "I'll love you *if* you come with me to visit my parents this Christmas."

- "I'll love you *if* you lose weight."

Conditional acceptance is based on your behavior rather than your being.

When we receive conditional acceptance, the message is that our worth is based on our behavior, that we are not innately worthy human beings. We consequently learn to displace our worth onto our behavior. Our worth then may become dependent on getting straight A's, winning the race, getting the job, or being valedictorian. The problem is not that these are invalid goals. It is the implicit message that *if we don't* get straight A's, win the race, get the job, *then we are unworthy.*

❖ EXERCISE 11:   CONDITIONAL ACCEPTANCE

Can you identify times when you received conditional acceptance from others? From yourself?

Can you identify times when you received unconditional acceptance from others? From yourself?

What effect did each have on you?

GENDER-ROLE EXPECTATIONS

Our desire for love, acceptance, and approval can result in our focusing on others' expectations of us. When we believe our worth is dependent on external judgments, we try to live up to those expectations.

There is a wide range of societal and cultural expectations that create external standards for judgment and approval. Influences such as family, school, church, peers, and media set up expectations for human behavior. The individual is bombarded with covert and overt messages from these sources regarding how to act. Nowhere are these expectations more evident than in gender roles.

> *Baby boys are described as strong, solid, and independent, and
> baby girls as loving, cute, and sweet.*   (PEARSON, 1985, P. 38)

Social scientists have long been concerned about the effect that gender-role conditioning can have on psychological adjustment and mental health. They have voiced particular concern about the potentially restrictive and delimiting effect that gender roles can have on individuals.

❖ EXERCISE 12:   SELF-DESCRIPTION

Consider the following descriptors. On a scale of 1 (not at all) to 7 (very), rate how descriptive each adjective is of you.

1.	emotional	1 2 3 4 5 6 7	11.	rational	1 2 3 4 5 6 7
2.	cooperative	1 2 3 4 5 6 7	12.	competitive	1 2 3 4 5 6 7
3.	dependent	1 2 3 4 5 6 7	13.	independent	1 2 3 4 5 6 7
4.	passive	1 2 3 4 5 6 7	14.	aggressive	1 2 3 4 5 6 7
5.	following	1 2 3 4 5 6 7	15.	leading	1 2 3 4 5 6 7
6.	relationship-focused	1 2 3 4 5 6 7	16.	career-focused	1 2 3 4 5 6 7
7.	submissive	1 2 3 4 5 6 7	17.	dominant	1 2 3 4 5 6 7
8.	weak	1 2 3 4 5 6 7	18.	strong	1 2 3 4 5 6 7
9.	sensitive	1 2 3 4 5 6 7	19.	insensitive	1 2 3 4 5 6 7
10.	expressive	1 2 3 4 5 6 7	20.	inexpressive	1 2 3 4 5 6 7

Add up your score for adjectives 1–10 in the first column, and for adjectives 11–20 in the second. Column 1 represents characteristics traditionally considered to be more appropriate for females and column 2 represents characteristics traditionally considered more appropriate for males (Bem, 1974; Spence, Helmreich, & Stapp, 1975). Consider your scores. What effect have gender-role expectations had on how you view yourself?

Historically, the acquisition of appropriate sex-typed behaviors, which resulted in a masculine identity in men and a feminine identity in women, was considered a necessary learning for mental health (Bandura, 1969; Broverman, Broverman, Clarkson, Rosenkrantz, & Vogel, 1970; Mischel, 1966). Traditionally, males have been expected to be strong, rational, responsible, aggressive, competent, competitive, dominant, independent, able to lead, and career focused; females have been expected to be sensitive, emotional, submissive, nurturing, cooperative, passive, dependent, understanding, care taking, content to follow, and relationship focused (Bem, 1974). Although masculine and feminine traits are no longer viewed as mutually exclusive bipolar opposites (Chafetz, 1988; Long, 1986, 1989), gender role conditioning has been salient and powerful.

The Masculine Gender Role and Self-Concept

The socialization process for males has stressed competency-based, instrumental attributes and has in many ways been more stringent than the socialization process for females. A "tomboy" girl, for instance, has generally been more accepted than a "sissy" boy. Studies have shown that boys receive significantly more disapproval for cross-sex behavior than do girls (Gayton, Sawyer, Baird, & Ozman, 1982).

> *Men are especially likely to be judged negatively for behaving in feminine ways in our culture.* (PEARSON, 1985, P. 46)

Traditional gender-role expectations that have the potential for restrictive effects on males have been summarized by Pleck (1983) as including what can be referred to as the three P's: performing, providing, and protecting.

For evidence of traditional expectations of male performance, we are referred to the athletic field, the work place, and the bedroom. The expectation that men provide financial and economic security has been historically consistent. Although more and more women work outside the home, their adequacy as women has not traditionally been based on the job they hold, and for women not to work is still considered socially acceptable.

> *A man's socialization is still to be able to provide for a woman as well as himself.* (FARRELL, 1986, P. 11).

Evidence of traditional expectations of males as protectors is seen in every war and every time a female defers to a male to investigate a strange noise outside the house or in the basement.

❖ EXERCISE 13:   EXPERIENCES OF THE THREE P'S FOR MALES

Identify times when you have experienced the three P's if you're male, and times when you've encouraged or observed the three P's if you're female.

What effect does the expectation of the three P's have on your view of yourself?

The Feminine Gender Role and Self-Concept

The socialization process for females has historically been toward relationship-oriented attributes. Traditional gender-role expectations that have the potential for restrictive effects on females can be viewed as including the three D's: devalue, defer, and depend.

The devaluation of females can be seen when women are expected to be less capable and competent than men, when women are paid less than men for the same work (Schau & Heyward, 1987), and when women are underrepresented in government, administration, and positions of power.

> *Studies on the competence of men and women have shown that women are generally judged to be less competent than men.* (PEARSON, 1985, P. 49)

> *Women live in a culture where their perceptions are rarely validated and they are taught from birth that men are innately superior.* (SCHAEF, 1985, P. 53)

Females defer to males. Studies have shown, for example, that in mixed groups, the topic of conversation is more likely to be determined by males, that males are more often allowed to interrupt, and that credit is more often given to males for their comments than to females (Association of American Colleges, 1986; Tannen, 1990). One study by Emily Hancock (1981) showed that women viewed themselves as supporting actresses in their lives, deferring to their husband and children, who were playing the leading roles. Dependency can be encouraged in girls from a very young age. In elementary school, teachers have noted that boys can be anywhere in the room and still maintain contact with the teacher, but that girls need to be right next to the teacher to receive the same kind of attention. The notion that a girl has to sit next to the teacher encourages dependency in girls (Sadker & Sadker, 1981).

❖ **EXERCISE 14:   EXPERIENCES OF THE THREE D'S FOR FEMALES**

Identify times when you have experienced the three D's if you're female, and times you've observed or encouraged the three D's if you're male.

What effect does the expectation of the three D's have on your view of yourself?

❖ **EXERCISE 15:   CHALLENGES RELATIVE TO GENDER**

What challenges do you see for yourself, relative to gender conditioning and expectations?

A major trauma for me, which began in high school and lasted through college, came when my peers started dating. While I felt compelled to at least try to be successful at being a female, that seemed particularly difficult to accomplish when my interests were primarily in the horse barn, mountains, and woods. So, although I dutifully scheduled social events and encounters in my calendar, I felt out of sync with myself, and lost in my own life.

CHAPTER 4

Resulting Issues:
The Inherent Dilemma,
Discrepant Self, and Lost Identity

THE INHERENT DILEMMA

Externally Based Happiness and Worth

The experience of focusing externally for judgment and approval can result in our developing the belief that our happiness and worth are dependent on the external world. We begin to believe that others know what's best for us, that happiness comes from outside ourselves, and that our *worth* is based on external approval. As a result, we can begin to feel lost in our own lives. Following are potential external beliefs—beliefs that our happiness and worth have a basis outside ourselves—with illustrating examples.

- Belief 1.   Our happiness is dependent on external events.

  Our happiness is dependent on aquiring a new car, getting married, getting divorced, securing a better job, or moving to a bigger house.

- Belief 2.   Our worth is dependent on what we do, accomplish, and achieve rather than who we are.

  Our worth is dependent on making straight A's, getting a promotion, winning a community service award, or hitting a home run.

- Belief 3.   Our happiness and worth are based on others' approval rather than our own approval.

  Our worth is dependent on a friend's acceptance, a partner's love, a parent's approval.

- Belief 4.   Our happiness is relative to others' happiness.

  Our happiness is dependent on our making a better grade than someone else, on our children's doing better than others' children, or on having a nicer house than someone else has.

We sometimes mistakenly believe that "happiness will result from being better off than others are."   (SMITH, DIENER, & DOUGLAS, 1989, P. 319)

❖ **EXERCISE 16:   EXTERNAL BELIEFS**

Identify examples that reflect the extent to which you believe that your happiness and worth are dependent on your external world.

1.   *My happiness and worth are dependent on external events.*

2.   *My worth is dependent on what I do, accomplish, and achieve.*

3.   *My worth is based on others' approval of me.*

4.   *My happiness is relative to others' happiness.*

Problems with External versus Internal Judgment

Focusing externally for acceptance and approval is initially a natural part of human growth and development. Our initial self-concept is formed through external feedback. We have a basic human need for love and acceptance (Maslow, 1968). We learn social and cultural norms for behavior through observation, role modeling, and social reinforcement. Focusing externally plays an important role in our development. Ideally, we learn self-acceptance and approval through the experience of receiving acceptance and approval from others, as well as by observing others demonstrating self-acceptance. The external feedback we receive in our initial self-concept formation may not, however, be ideal.

Focusing on external sources for acceptance and approval can create a potential problem. The problem arises when external focus for approval interferes with the development of a positive self-concept. When we equate our worth with external approval, our self-concepts are dependent on others' judgment. By definition, a positive self-concept is based on the individual's *own* judgment and approval. If our worth, therefore, is dependent on external approval, we cannot, by definition, have a positive self-concept.

THE INHERENT DILEMMA

By nature, the human being wants love and acceptance. We initially perceive our worth to be dependent on the love, acceptance, and approval of others. The power of judgment is initially *external*.

A positive self-concept is one the *individual* judges to be positive. The power of judgment is *internal*.

To develop a positive self-concept, then, we must focus on internal judgment. We are, however, more practiced at focusing externally for judgment. This, then, is the inherent dilemma in the development of a positive self-concept: We develop the habit of equating our happiness and worth with *external* events, judgment, and approval. A positive self-concept, by definition, however, is based on *internal* judgment and approval.

❖ EXERCISE 17:   THE INHERENT DILEMMA

To what extent do you equate your happiness and worth with external judgment and approval?

To what extent do you believe happiness and worth are experienced internally?

THE DISCREPANT SELF

In wanting acceptance and approval, we initially focus externally on cues and responses from others, trying to adjust who we are to gain acceptance. In trying to adjust who we are to meet these external expectations, we can develop a discrepant self: an "external self," developed in response to external expectations, one that is discrepant from our experienced "internal self." The discrepant self has been referred to as the ideal versus the real (Perls, 1969), the artificial versus the actual (Heuscher, 1992), the self-image versus the experienced one (Rogers, 1980), and the perceived versus the actual self (Woody, Hansen, & Rossberg 1989). Regardless of terminology, there is consistency in the assessment that development of a discrepant self has negative impact on the growth of a positive self-concept.

External versus Internal Selves

The external self is the image we try to present to the world to gain acceptance and approval. For the external self, worth is dependent on meeting certain conditions. The conditions (getting straight A's, marrying, becoming a doctor) are often initially determined by external sources (parents, peers, media) and later internalized. When the conditions are met, you feel worthy. The feeling of worth is temporary, however, because the conditions need to be met continually. When the conditions are unmet, you experience a sense of unworthiness.

The internal self is our real, actual experience of ourselves. To the extent that you have a discrepant self, your internal self may feel as though it is hiding behind the external self. The external self can feel like a "mask" the internal self is wearing and presenting to the world.

❖ EXERCISE 18:   THE DISCREPANT SELF

Draw a mask on the next page that reflects your external self. Write at least ten things on the mask that reflect ways you think you have to be to get approval. These represent how you believe you should be or what you should do.

Now draw a face that symbolizes your actual experience of yourself—your internal self. Write ten things that reflect your internal self.

Describe the discrepancy between your external and internal selves.

External Self-Defense

Once we have formed these masks, we tend to maintain and defend them for several reasons. In part, it is simply a familiar habit. Once a habit is formed, we sometimes don't question it again. The familiar feels comfortable, and changing it is disruptive. Additionally, we may defend the known out of fear of the unknown. The familiar feels comfortable; the unfamiliar seems scary. If we don't keep the mask we have, what will we have in its place? Finally, we defend our masks because of a genuine belief that our very survival is at stake. If our masks are all we know ourselves to be, and we have not discovered our true self-identities, we don't know what the masks are concealing. Defending them is defending our very perceived existence.

We learn how to defend our [mask] images from attack, whether this be assault from without or doubts from within. Almost every elementary psychology text contains a long list of "defense mechanisms," which attest to the ingenuity we summon in defense of that image. (ARKOFF, 1988, P. 9)

❖ **EXERCISE 19:   DEFENSES**

What defenses do you use to maintain your mask? In what ways does it reflect habit, fear of the unknown, and/or perceived survival?

We present our external selves to others because we believe they are what others will approve, and we want their approval. Ironically, even if we get the desired acceptance and approval, we then don't feel it, because we know it's our external selves—our masks—that are getting the approval. *We*—our real internal selves—are behind the masks, watching and still wanting acceptance and approval!

LOST IDENTITY

We originally seek approval from others as a means to an end. We want approval in order to feel good about and approve of ourselves. Our goal is actually *internal* acceptance and approval. If we lose sight of our real goal, however, approval from others becomes the end in itself (Cushman, 1990), with *external* acceptance and approval replacing the self-approval we really want.

> *We can get so distracted by maintaining an external image that the image takes priority over our being; for example, our image of ourselves as parents can become more important than being good parents.*   (ADAPTED FROM COVEY, 1990, P. 98)

We can become so focused on developing an external self to acquire external approval that we don't take the time to discover who we really are. We don't develop our internal selves—our true beings.

> *A major problem for many people is that they have lost a sense of self because they have directed their search for identity outside themselves. "Each of us would like to discover a self—that is, to find (or create) our personal identity.... The trouble with so many of us is that we have sought directions, answers, values, and beliefs from [others]... rather than trusting ourselves....Our being becomes rooted in their being, and we become strangers to ourselves."*   (COREY, 1991, P. 180)

> *Rollo May (1989) suggests that too many of us have become "hollow people" who have very little understanding of who we are and what we feel, resulting in an "inner emptiness." He cites one person's succinct description: "I'm just a collection of mirrors, reflecting what everyone expects of me."*   (P. 15)

> *We become so adept at masking [ourselves]...hiding...from ourselves as well as from others—[that] we become strangers to ourselves and scarcely know who we are.*   (ARKOFF, 1988, P. 4)

❖ **EXERCISE 20:   LOST IDENTITY**

To what extent do you feel you have lost your own identity?

In Search of the Self

To have a positive self-concept you must first have a self to conceptualize as positive. Perhaps the gravest potential consequence of the discrepant self is the loss of self-identity. Discovering our self-identities involves bringing our discrepant selves into focus.

❖ EXERCISE 21:   IN SEARCH OF SELF: BRINGING OUR DISCREPANT SELVES
INTO FOCUS

Examine your masks from the previous exercise. Look at the "shoulds" that you wrote. List each "should" below and determine its source. For each one ask yourself if it is something you really want or if it is something that someone else wants for you. If it's something you want, list it as a want. If it's not something you want, erase it from your list. The list will then reflect the goals of who you want to be.

THE SEASONS OF FREEDOM

I was born to the playground
in summer.
To soft sky, warm sun
and good earth

I was free to move
with the breezes.
Free of fear, clothing
and girth.

I was still on the playground
in autumn.
Though for fear I'd be harmed
I was clothed.

I wandered restricted
and wondered
at the need for constriction
I loathed.

I stood on the playground
in winter.
Deep under layers of snowsuit
and wools.

My soul cried out
in anger
As I peered out at the
playground through holes.

The playground in spring
appeared dampened
When they came
To remove my coat.

They said I should play
but I couldn't
I felt stiff and withered
and cold.

Now the playground again
is in summer
I look, lightly clad,
and grown,

And remember the movement
of breezes
and begin to recall
my own.

—VONDA LONG

Personal Growth Model: Developing a Positive Self-Concept

A positive *self-concept* means feeling satisfied with who you are and how you're living your life.

When I experience fear and anxiety over lack of acceptance by others, I am saying that their judgment is more important than my own.

CHAPTER 5

Components of a
Positive Self-Concept

A positive self-concept has been defined as feeling satisfied with who you are (your essence, your being), and how you are living your life (your choices, actions, living). This definition implies that individuals make internal assessments and judge independently whether they are happy with themselves and their lives. Individuals determine whether and to what extent they have a problem or an issue. Since individuals make their own judgment of themselves, implicit in a positive self-concept is respect for their own diverse differences in culture, ethnicity, spirituality, sexual orientation, ability, education, socioeconomic status, and family background.

A positive self-concept is a conceptualization of yourself— including the human dimensions of affective experience (being), behavior (doing), and cognition (choosing) which, in your judgment is positive.

❖ **EXERCISE 22:   DEVELOPING A POSITIVE SELF-CONCEPT**

Assuming you would like to develop a more positive self-concept, what would you identify as the components to be developed?

171

A positive self-concept involves feeling satisfied with yourself and how you're living your life. This necessarily incorporates all our human dimensions. The components of a positive self-concept, therefore, might be considered by looking at how individuals experience themselves in each of their human dimensions: affective, behavioral, and cognitive.

AFFECTIVE DIMENSION

Affective experience influences how you feel about yourself. It reflects the ultimate outcome goal: feeling satisfied with who you are and how you are living your life. Components of a positive self-concept in the affective experience dimension include these:

- Self-acceptance—valuing yourself despite perceived shortcomings; accepting your *being*

- Self-esteem—valuing yourself based on perceived strengths; esteeming your *doing*

- Self-actualization—valuing yourself based on your choices and how you're living your life; actualizing yourself through your *choosing*

Although it is difficult to observe and measure components of the affective dimension, they are reflected more concretely in the components of the behavioral dimension (see Table 5.1).

BEHAVIORAL DIMENSION

The behavioral dimension focuses on your actions and behavior. If you are feeling good about yourself and how you're living your life, your assessment would be reflected in your behavior. Components of a positive self-concept in the behavioral dimension include the following:

- Congruence—Consistency of thoughts, feelings, and behavior; a *being* that reflects an integrated rather than discrepant self; self-acceptance

- Competence—Capability reflected in actions; *doing* that results in self-esteem

- Internal Control—Decisions and *choices* that reflect self-actualization

Whereas behavioral components reflect your affective experience, behaviors are based on your underlying beliefs. Each behavioral component, therefore, reflects an underlying belief from the cognitive dimension (see Table 5.1).

COGNITIVE DIMENSION

The cognitive dimension includes your thoughts, underlying beliefs, and values. These components are reflected in your behavior and influence your affective experience. Components of a positive self-concept in the cognitive dimension include the following underlying beliefs:

- Rights—The belief that individuals have the right to be who they are—to have their own thoughts, feelings, and behaviors (short of harm to others). If you believe you have the right to be who you are, it is reflected in your *congruence*, and you experience *self-acceptance*.

- Respect—The belief that individuals, with all their unique differences, are worthy of positive regard and are capable and able. Self-respect is reflected in *competence*, and you experience *self-esteem*.

- Responsibility—The belief that individuals are accountable for their actions and choices. Responsibility is reflected in *internalized control*, and you experience *self-actualization*.

Growth can be defined as movement toward improved psychological health through the development of a positive self-concept. One way to approach developing a positive self-concept is to facilitate the development of its components. The affective experience components of self-acceptance, self-esteem, and self-actualization are therefore addressed in Chapter 6. Developing the behavioral components of congruence, competence, and internal control are discussed in Chapter 7. The underlying beliefs concerning rights, respect, and responsibility—the cognitive components—are examined in Part III.

Human Dimension:	BEING	DOING	CHOOSING
Affective Experience Dimension	Self-Acceptance	Self-Esteem	Self-Actualization
Behavioral Dimension	Congruence	Competence	Internal Control
Cognitive Dimension	Rights	Respect	Responsibility

TABLE 5.1
Components of a Positive Self-Concept

> *When I allow myself to be totally in tune with this moment then I find myself more ready to deal with the next one.*

CHAPTER 6

A Positive Self-Concept:
Self-Acceptance, Self-Esteem, and
Self-Actualization

Components of a positive self-concept for the affective, behavioral, and cognitive human dimensions are shown in Table 5.1. The three components of affective experience—self-acceptance, self-esteem, and self-actualization—however, can be described as the ultimate outcome goals of a positive self-concept. They are the desired outcomes to which we aspire. They reflect the very definition of a positive self-concept: feeling satisfied with who you are and how you're living your life. An individual aspiring to develop a positive self-concept would find it useful to assess his or her self-acceptance, self-esteem, and self-actualization.

SELF-ACCEPTANCE

Self-acceptance is unconditional acceptance of the individual's inherent worth, given regardless of perceived shortcomings.

Self-acceptance involves self-identity and unconditional acceptance. Self-identity is *knowing* who you are. Unconditional acceptance is *accepting* who you are once you know.

Self-Identity

It's hard to *accept* who you are before you *know* who you are. Self-identity means bringing our discrepant selves into focus and recognizing our internal selves as worthy. It means acknowledging our core essence, which is constant, as unconditionally worthy, and being able to separate it from our thoughts, feelings, and behaviors, which are constantly changing.

175

Self-identity is a prerequisite to self-esteem and self-actualization as well as self-acceptance. It's hard to esteem or actualize yourself without *knowing* yourself. Indeed, self-identity is a prerequisite to a positive self-concept. It's hard to conceptualize as positive something you don't know.

❖ EXERCISE 23: SELF-IDENTITY

Describe the essence of your self-identity.

Unconditional Acceptance

Acceptance does not mean approval, consent, permission, agreement, or endorsement. You don't have to like what you accept; you simply accept it. For example, I can *accept* my having straight hair, curly hair, diabetes, artistic talent, no musical talent, or a certain ancestry, without necessarily *liking* it. Self-acceptance is also *not* the same as passivity or inaction. Just because I *accept* myself does not mean I am not actively working toward improving myself. For example, I might *accept* being a slow reader; at the same time, I might be taking a speed-reading course to work on improving my speed. Indeed, until we *accept* the reality of what is, moving beyond it will be difficult for us.

❖ EXERCISE 24: SELF-ACCEPTANCE

Assess your self-acceptance by identifying ten perceived shortcomings or weaknesses that you can view with acceptance, still feeling good about yourself.

1.

2.

3.

4.

5.

6.

7.

8.

9.

10.

Obstacles

The lack of self-identity, prioritizing of achievements, and development of an external self, or "mask," can create major obstacles to self-acceptance.

❖ EXERCISE 25:   OBSTACLES TO SELF-ACCEPTANCE

Do a self-assessment of the obstacles to your self-acceptance.

SELF-ESTEEM

Self-esteem is self-respect earned by the individual for perceived strengths, attributes, and actions.

Self-*acceptance* is given for inherent worth despite what we do; self-*esteem* is earned by what we do. Self-esteem is based on actions and perceived strengths. It is feeling satisfied with how you're living your life. Actions are not the same as outcomes. Self-esteem, therefore, is based on how you play, not whether you win. Self-esteem means the individual acts with integrity and lives life in a way that he or she respects and esteems. There is no substitute for self-esteem.

❖ EXERCISE 26:   SELF-ESTEEM

Assess your self-esteem by listing ten perceived strengths (abilities, actions, competencies) that you feel good about.

1.

2.

3.

4.

5.

6.

7.

8.

9.

10.

Obstacles

False self-esteem is based on others' shortcomings rather than our own strengths. Blaming others rather than developing our own competencies and abilities, and using achievements and accomplishments as a way of trying to *prove* our worth are examples of obstacles to developing real self-esteem.

❖ **Exercise 27:   Obstacles to Self-Esteem**

Do a self-assessment of the obstacles to your self-esteem.

SELF-ACTUALIZATION

> *Self-actualization means the individual is actualizing—or becoming and being—himself or herself. The person is allowing his or her innate potential to flourish and the true identity to emerge.*

Self-actualization involves recognizing your internal self and choosing to actualize yourself. It is the process of becoming and being who you are. It is a process of self-discovery of your true identity, and an actualizing of your innate potential.

❖ **Exercise 28:   Self-Actualization**

Only you know whether you are being true to yourself. Only you know whether you are living your life in a way that makes you feel good. Only you know whether you are actualizing yourself, because only you know who you are. Identify recent choices you have made that reflect your actualizing who you are and being true to yourself.

Identify recent choices you have made that do not reflect your actualizing yourself and being true to yourself.

What influenced you in making decisions in both cases? How did you feel about each set of decisions?

Obstacles

Potential obstacles to actualizing ourselves include unmet basic needs such as security, love, and acceptance; the need for external approval; and the belief that our worth, happiness, and well-being are dependent on others.

❖ EXERCISE 29:   OBSTACLES TO SELF-ACTUALIZATION

Do a self-assessment of the obstacles to your self-actualization.

I was heavily influenced by the concept of saving—saving money, saving the good dishes for company, saving linen and towels for my hope chest, saving the best for somebody else, and saving the present for some future time. I remember one time when I was a little girl, saving a Busy Baker Cake-N-Frosting mix until it dried out and disintegrated, and I had to throw it away. Later, I remember getting a coat that I really liked, and saving it for special occasions. There were so few occasions that were special enough to wear it, that it hung in my closet unworn. Years later, when it was completely out of style but still brand new—I wore it around doing chores. There was a blatant irony in that experience. I realized that enjoying something in the present—be it a coat, or cake mix, or time—is the single best guarantee that it will be enjoyed.

A Positive Self-Concept: Developing Congruence, Competence, and Internal Control

Although self-acceptance, self-esteem, and self-actualization are what we ultimately aspire toward, they are difficult to observe or develop directly. They are, after all, how we *experience* ourselves.

We can *infer* self-acceptance, self-esteem, and self-actualization, however, by observing our behavior. Self-acceptance may be inferred by observing congruence, self-esteem by observing competence, and self-actualization by observing internal control (see Table 5.1). Congruence, competence, and internal control, then, can be viewed as components of the behavioral dimension that reflect our ultimate goals of self-acceptance, self-esteem, and self-actualization. We propose that one way to promote the development of a positive self-concept is to develop congruence, competence, and internal control.

CONGRUENCE

Congruence means the "quality or state of agreeing, corresponding, or coinciding" (*Webster's New Collegiate Dictionary*, 1981). In the context of a positive self-concept, congruence means that you are yourself. The external self is coinciding, or congruent, with the internal self. You are consistent with yourself in all your human dimensions—thoughts, feelings, and actions. Congruence requires giving yourself permission to be yourself in the moment, and it allows you to be fully present and in the "now." Congruence is a prerequisite to actually participating in and enjoying the process.

Promotion

Congruence can be promoted by
1. Integrating external and internal selves
2. Encouraging consistency of thoughts, feelings, and actions

3.   Focusing on being present in the process and living in the moment

❖ **EXERCISE 30:   DEVELOPING CONGRUENCE: INTEGRATING EXTERNAL AND INTERNAL SELVES**

In a previous exercise you identified the "shoulds" of your external self or "mask." You then translated the "shoulds" you wanted to keep into "wants," and deleted the others.

Below, describe yourself first by writing the traits and characteristics of yourself that you feel good about (e.g., artistic, friendly, intelligent). Then add the desired "wants" as characteristics you're aspiring toward. Put asterisks (*) by the aspirations. This description reflects the integration of your external and internal selves.

When we are being congruent, we are accepting ourselves—our thoughts, feelings, and behavior. Accepting is *not* necessarily the same as liking or condoning. When we are not congruent, we are not accepting a part of ourselves. This does *not* mean we must always elect to share or act on all our thoughts, feelings, and behaviors. It means we need to acknowledge and accept them ourselves. Only *we* know when we are rejecting a part of ourselves.

❖ **EXERCISE 31:   DEVELOPING CONGRUENCE: ENCOURAGING**
                CONSISTENCY OF THOUGHTS, FEELINGS,
                AND BEHAVIOR

Identify a time when you wanted to share your thoughts, beliefs, or opinions, but didn't do so for fear of being rejected, and as a result you rejected a part of yourself.

Identify a time when you didn't share your feelings for fear of being rejected, and as a result you rejected a part of yourself.

Identify a time when your actions did not reflect what you wanted to do, and as a result you rejected a part of yourself.

How did you feel about yourself in these examples? In each case, what could you have done in order to have been congruent?

Sometimes we can get so preoccupied with the task or the outcome that we forget to enjoy the process. We attempt to earn A's at the expense of enjoying education. We plan vacations and parties and become so preoccupied with accomplishing them that we don't have any fun. Enjoying the process is separate from enjoying the outcome. This doesn't mean we can't have goals, or desire particular outcomes. It simply means that our enjoyment of the process is not *dependent* on the outcome.

❖ **EXERCISE 32:**   **DEVELOPING CONGRUENCE: FOCUSING ON BEING PRESENT IN THE PROCESS AND LIVING IN THE MOMENT**

Identify an experience when you were so focused on the outcome you didn't enjoy the process.

Identify an experience when you really enjoyed the process, even though the outcome was not what you expected.

I remember spending a week in New York, in an intensive personal growth group. I held back. I never shared my real self with the group, only selected and edited parts. The group accepted my selected parts. But I always wondered afterward if they would have accepted all of me. I'll never know because I never revealed myself. It wasn't until the end of the time with the group that I looked back and realized my chance had gone and that I wished I had given myself permission to have really been there.

Obstacles

A primary obstacle to developing congruence is the fear of what might happen to us if we let ourselves be who we are. This fear is primarily a fear of rejection. Ironically, by not letting ourselves be who we are, *we* reject ourselves, before others have a chance to do so.

❖ EXERCISE 33: OBSTACLES TO CONGRUENCE

Do a self-assessment of the obstacles interfering with your congruence. What are you afraid would happen if you let yourself be who you are?

COMPETENCE

Competence means "capable, able, adequate, sufficient, or having the qualities necessary for..." (*Webster's New Collegiate Dictionary*, 1981). In the context of a positive self-concept, competence means being capable of living your life in a way that satisfies you. It means you develop and use your qualities, capabilities, and abilities in ways to enhance your personal integrity. It is living your life using all your human dimensions in a way that you respect and feel good about. Competence involves the development and implementation of capabilities, abilities, and skills. It involves integrity of actions.

There is no substitute for competence. Competence means the individual is capable and able to do what he or she needs to do. Perhaps, for example, you feel as though you let a friend walk all over you. You feel the friend took advantage of your good nature and convinced you to do volunteer work that you didn't want to do. You weren't able to stand up for yourself and say "No." Having someone else go talk to the friend and say that you're not going to do it might solve the external problem, but not the internal problem: feeling unable to stand up for yourself. If you play in a recital and you don't feel as though you played well, no amount of consolation can substitute for the satisfaction you would have gotten from a good performance. There is no substitute for the *internal* feeling of satisfaction that comes from acting with competence.

Promotion

Competence can be promoted by

1. Identification of desired competencies
2. Action
3. Reframing mistakes

❖ EXERCISE 34:   DEVELOPING COMPETENCIES: IDENTIFICATION OF
                   DESIRED COMPETENCIES

Competencies can fall into three general categories: work, relationships, and self. Below, to the left, are examples of possible competencies in each area. In the middle column do a self-assessment of your own competencies as you perceive them. Then in the right column, identify competencies you would like to develop in each area.

	Self-Assessment	Desired Competencies
Work		
I am a capable teacher	well organized	computer skills
I'm not doing all that I'm capable of		
I'm a good writer		
I lack computer skills		
I am good at follow-through		
Relationship		
I am able to be intimate		
I have good social skills		
I relate well to others		
I am assertive		
I am capable sexually		
I am able to express feelings		
Self		
I can handle my emotions		
I am physically fit		
I like who I am		
I like how I behave		
I have abilities		

```
I am athletic
I am spiritually at peace
I am self-reliant
```

Without action, competence is only potential. Competence can not be developed without *doing* something. Competence requires action.

❖ EXERCISE 35:   DEVELOPING COMPETENCE: ACTION

Identify one desired competency from each area in the previous exercise, and identify a plan to begin developing it. Include a time frame for your plan.

Fear of failure or making mistakes often keeps us immobilized and prevents us from developing competencies. To the extent that we believe our worth is dependent on what we do, on our achievements and outcomes, we fear making mistakes. The fear is that mistakes reflect unworthiness. To the extent that we are able to recognize our inherent worth and differentiate that from our behavior, we can see mistakes for what they are. Mistakes are simply natural steps of the learning process.

❖ EXERCISE 36:   DEVELOPING COMPETENCE: REFRAMING MISTAKES

Identify a time when you associated making a mistake with your worth as a human being.

Identify a time when you avoided doing something (e.g., learning a sport), to avoid making a mistake.

Obstacles

A primary obstacle to competence is the fear of what we might find out about ourselves if we actually try. That fear is primarily a fear of failure. Ironically, if we don't try, we not only get no practice building the skills we want, but we never get to experience the success we need to reinforce a belief in our competence. We automatically fail.

> *Whatever I do is going to be "good" if my criteria for "good" are that I'm enjoying doing it—and enjoying who I am in the process.*

❖ EXERCISE 37:   OBSTACLES TO COMPETENCE

Do a self-assessment of the obstacles interfering with your competence. What are you afraid would happen if you tried to develop the competencies you desire?

INTERNAL CONTROL

Control means "to direct, regulate, or exercise authority or power over" (*Webster's New Collegiate Dictionary*, 1981). In the context of the self-concept, internal control means an internal regulation of power over yourself. It implies having a choice, making decisions, and *choosing*.

Actualizing the self requires recognizing yourself and choosing to be true to yourself. A positive self-concept involves making choices to live in such a way that you feel good about yourself and how you're living your life. Only *you* can know whether *you* are satisfied with who you are and how you're living. Only you can know whether you are happy. Consequently, you alone can take responsibility for your own happiness, because when you actualize yourself, no one else can effectively choose for you.

Internal control is a power of choice that carries with it both the freedom and the right to choose, and the burden of responsibility for one's choices.

Promotion

Internal control can be promoted by

1.   Making conscious choices
2.   Being accountable for your actions
3.   Owning your feelings

I learned from my father that if, on my deathbed, I looked back on my life and was not happy with the way that I lived it, I would have no one to blame but myself. This may initially sound a bit harsh, but it's also permission giving. Because only *you* know whether you're living your life in a way that satisfies you, only *you* can make the choices regarding how to live it.

❖ **EXERCISE 38:   DEVELOPING INTERNAL CONTROL: MAKING CONSCIOUS CHOICES**

Imagine yourself on your deathbed. Write an obituary describing yourself and your life. Write it in a way that accurately reflects your life as it is.

Now rewrite your obituary imagining in retrospect, who you'd like to have been and how you'd like to have lived your life.

To what degree are you living your life as you want to live it? If you are not living your life as you'd like to, why not? What's keeping you from choosing to be who you want to be, and choosing to live in a way you'd feel good about?

Identify five-year goals, one-year goals, and immediate goals to help you begin living your life in a way you would feel good about.

Unless we are physically overpowered and forced, our actions reflect our choices. We choose our behavior. What we can assess is why we choose a particular behavior (e.g., were we motivated by fear?) and whether we would prefer to choose a different behavior.

❖ **EXERCISE 39:**   **DEVELOPING INTERNAL CONTROL: BEING ACCOUNTABLE FOR YOUR ACTIONS**

Identify a time you did something you didn't want to do. What was your motivation? What were your choices? What would be your preferred choice of behavior?

We sometimes try to convince ourselves that we are not responsible for our feelings. "You're making me angry," "You hurt my feelings," and "You're scaring me" all imply that the other person is *making* you feel the way you do. The reality is that we *choose* our feelings. We can see this clearly when we observe two or more persons responding to the same circumstance with different feelings.

After recess was over, one fourth grader hid behind the door and playfully leaped out as a group of girls came past. One girl shrieked, one laughed, and one cried, reflecting the range of their reactions.

❖ **EXERCISE 40:**   **DEVELOPING INTERNAL CONTROL: OWNING YOUR FEELINGS**

Identify a situation in which you chose a feeling you didn't like, such as anger, jealousy, hurt. What feeling would you prefer to have felt? What kept you from choosing the preferred feeling?

Obstacles

A primary obstacle to internalized control is the fear of the unknown. That fear is primarily a fear of abandonment. We are in the habit of focusing externally and having others decide how we should be and how we should live our lives. To focus internally for such decisions is to pursue an unknown, to leave the familiar behind us. We then feel as though we have been abandoned by the familiar. Ironically, by not allowing ourselves to choose ourselves, we are abandoning ourselves.

Yesterday I took the afternoon off from work and came home to write. I never did write— I read a novel instead. While I was reading I periodically thought that I should be writing It wasn't until the afternoon was over that I looked back and, in retrospect, thought that as long as I was going to spend the whole afternoon reading, I wished I had given myself permission to abandon thoughts of what I thought I should be doing and let myself really enjoy what I was doing

❖ EXERCISE 41:   OBSTACLES TO INTERNAL CONTROL

Do a self-assessment of the obstacles interfering with your internal control. What do you think would happen if you began taking conscious responsibility for your choices, feelings, and actions?

ON LIFE AND LOVE

Golden Sun
And salty breeze
Softly light
The rhythmic sea.

Through troughs and swells
I gaily rode
As sea heaved sighs
And currents flowed.

Then caught by a crest
And toppled to shore
I sat stunned and crumpled
On the sandy floor.

But golden sun
And salty breeze
Touch the sea
And beckon me.

—VONDA LONG

Underlying Beliefs
of a Positive Self-Concept:
Rights, Respect, and Responsibility

Underlying beliefs affect our thoughts, feelings, and behaviors.

My focus on what I think I should be doing or accomplishing inevitably keeps me from getting in touch with what I want to be doing—or experiencing that which I am doing. Consequently I not only don't "get anything done" but I also keep myself from enjoying myself while I'm doing it.

CHAPTER 8

Underlying Beliefs:
The Foundation
for a Positive Self-Concept

UNDERLYING BELIEFS

What we believe affects how we think, feel, and behave. Generally, we experience events in our lives through a process of interpretation (thinking), emotional reaction (feeling), and behavior (acting). These processes, however, are affected by our underlying beliefs.

I had a client once whose boyfriend was going backpacking with a male friend without inviting her. Her behavior was to give him an ultimatum that either he take her or she would break up with him. That behavior was her reaction to feeling hurt. The feeling of hurt was a reaction to thoughts that included "I must not be a good enough hiker," or "He must not love me."

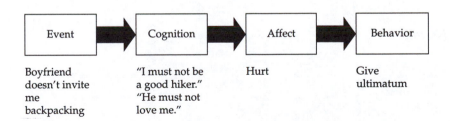

Event	Cognition	Affect	Behavior
Boyfriend doesn't invite me backpacking	"I must not be a good hiker." "He must not love me."	Hurt	Give ultimatum

Obviously, there must have been an underlying belief that affected how she viewed the event, causing her to interpret it as "He must not love me." Someone else may have interpreted it differently. In this case, the underlying belief might be "In order to be a worthy person, I must be good at everything," or "If I'm a worthy person, my boyfriend will want me to be with him all the time." The underlying belief affected her interpretation of the event, resulting in her feeling hurt.

If the underlying belief was replaced with one such as, "I'm a worthy person. I don't have to be good at everything," the interpretation of the event would have been different. It might have been interpreted as "My boyfriend wants to go back-packing. I don't really enjoy backpacking. I'm glad he has someone to go with so I don't have to feel obligated!"

❖ EXERCISE 42:   UNDERLYING BELIEFS

Identify a recent event that triggered a behavioral and/or emotional reaction for you. Identify the (1) thoughts, (2) feelings, (3) behavior, and (4) underlying belief that you used in response to this event:

Event

(Underlying belief)

thoughts

feelings

behavior

Identify underlying belief last.

Is the underlying belief one you want to keep? If not, what belief would you want to substitute?

Often our behavior is reactive. We *react* to the feelings created by our interpretation of an event. Our interpretation was based on our underlying beliefs. Reactive behavior is often inconsistent with the goal that we actually want. My client didn't even like backpacking and didn't really want to go. What she really wanted was to feel secure in her own worth. If she had felt more secure, her boyfriend's choosing to do something without her wouldn't have caused her to question his love for her and her own worthiness.

❖ EXERCISE 43:   REACTIVE BEHAVIOR

Identify an event in which your behavior was reactive. Did it get you what you wanted? How did you feel about yourself?

Event

(belief)

thoughts

feelings

reactive behavior

consequence (how you felt about yourself)

Now replace the reactive behavior with a proactive behavior (one that would be consistent with what you wanted). What effect do you think it would have on how you felt about yourself?

Event

*(belief)

thoughts

feelings

*Identify underlying belief last.

behavior

consequence (how you felt about yourself)

UNDERLYING BELIEFS REGARDING RIGHTS, RESPECT, AND RESPONSIBILITY: THE FOUNDATION FOR A POSITIVE SELF-CONCEPT

Self-acceptance, self-esteem, and self-actualization—the affective experience outcome goals of a positive self-concept—are reflected in congruence, competence, and internal control; these are influenced by underlying beliefs regarding rights, respect, and responsibility (see Table 5.1). The belief that you have the right to be yourself is a prerequisite to congruence. Self-respect for your capability and abilities affects your development of competence. Believing that you are responsible for your own actions and choices results in internal control. We propose that one way to help promote the development of a positive self-concept is to explore your underlying beliefs regarding rights, respect, and responsibility.

I have discovered that my happiness does not depend on what I am doing—or if I'm doing anything. What matters is that I'm enjoying being who I am while doing it.

CHAPTER 9

The Right to Be Yourself

Being yourself is a prerequisite to developing a positive self-concept because you must first be yourself in order to conceptualize yourself in a positive light. Self-acceptance—one component of a positive self-concept—is reflected by congruence; congruence is based on believing you have the right to be yourself.

Conceptual Belief 1:

Right to Be Yourself

Each of us is inherently worthy and has the right to be and become who we are, as long as we don't infringe on the rights of others in the process.

Often, we do not learn how to be true to ourselves. We learn instead how to develop an external self and present it as a mask to the world in order to gain acceptance and approval. There may be times, of course, when presenting aspects of our masks to the world may be appropriate. The problem arises when we discover that our worth is attached to our masks; we find that our masks are living our lives for us or that we are unsatisfied with our lives.

❖ **EXERCISE 44: THE RIGHT TO BE YOURSELF**

To what extent do you really believe you have a right to be yourself?

Identify an experience when you were so preoccupied with presenting your external self—your mask—that you didn't enjoy yourself. Identify a time when you—your internal self—were completely present and accepted. Compare the experiences in terms of your self-concept.

Implicit in the underlying belief of the right to be yourself is the right to your own feelings, beliefs, and choices in living your life.

❖ EXERCISE 45:   RIGHT TO FEELINGS

Identify a recent time when you had a strong feeling (e.g., mad, glad, sad, scared). To what extent did you believe you had a right to feel it? What beliefs or fears, if any, interfered with your believing you had a right to the feeling?

Try practicing the following affirmation:

"I have a right to be true to myself. I have a right to my feelings."

Feelings themselves are not good or bad. You have a choice of actions in response to them.

❖ EXERCISE 46:   RIGHT TO BELIEFS

Identify a recent example of a time you shared or didn't share a belief or opinion (e.g., a response to a minority or gender joke you found offensive, or a reaction to hearing a friend berate a mutual friend). To what extent did you feel you had a right to your opinion/belief? Did you express your opinion/ belief? If not, why not? What, if anything, interfered with your belief in your right to your opinion?

Try practicing the following affirmation:

> *"I have a right to be true to myself. I have a right to my own beliefs and opinions."*

Opinions and beliefs in themselves are not good or bad. You have a choice of whether you want to keep or revise the belief, and to decide what actions you take as a result of it. Your right to your opinion does not mean others need to agree with it.

❖ EXERCISE 47:   RIGHT TO CHOICES

Identify a recent example of a time you made a decision to act (e.g., to go or not go home for Christmas, to skip class, to get married). To what extent was your decision a reflection of being true to yourself? To what extent did you believe you had a right to make your own decision? What beliefs or fears, if any, interfered with your belief in your right to be true to yourself?

Try practicing the following affirmation:

> *"I have a right to be true to myself. I have a right to make choices that reflect who I am."*

This does not mean you have a right to harm others or to interfere with their rights. It does not mean you need to be insensitive to what others want or how they feel. It means you have a right to make choices that reflect who you are—and so does everyone else.

In my fear of failure, and anxiety about performing well I used to ask myself "What's the worst thing that could happen?" Of course the answer was "to fail" or "perform poorly." And once I convinced myself that I could survive the failure—I could go ahead with whatever I was attempting to do. I now feel that the worst thing that could happen is to not feel good about myself while doing whatever I am doing.

CHAPTER 10

Self-Respect

Self-esteem, one of the outcome goals of a positive self-concept, is based on a respect for yourself—experiencing yourself as competent, capable, and able. Respect means to "feel honor or esteem for, to hold in regard, to value" (*Webster's New Collegiate Dictionary*, 1981). Respect is based on a belief in worth and value. There is no substitute for self-approval because no one else can give it; it comes only from you. We often desperately seek a substitution through approval from others, but no amount of consolation or approval from someone else can make up for a lack of self-respect.

Conceptual Belief 2:

Respect for Capability and Worth	Each of us is capable, able, and inherently worthy of respect as a human being.

❖ **EXERCISE 48:   SELF-RESPECT**

On what do you base your self-respect? To what extent do you respect your own capabilities and abilities? What do you need to do to earn your own respect?

Self-respect refers to a basic underlying belief in your own capabilities and abilities. It means you believe in your ability to be and develop competence. We often develop rationalizations, or counterbeliefs, that interfere with our developing competencies, that keep us from trying, and that keep us immobilized by fear of failure.

❖ **EXERCISE 49:   RATIONALIZATIONS INTERFERING WITH SELF-RESPECT**

Identify rationalizations that keep you from developing competencies:

"I'm not good at that."   "I don't have any coordination."

acting and trying:

"I can't do that."   "I've never done that."   "I tried that once…"

Try practicing the following affirmation:

"I am capable and able. I can develop the competencies I need."

This does not mean you already have the abilities; it means you are capable of developing them.

Sometimes after I've experienced a negative consequence from a behavior, such as feeling uncomfortable from eating before bed, or find myself alone and feeling lonely on a particular night or weekend because I didn't want to commit myself to plans with someone, then I feel tempted to structure a rule: I'll never do that again. But if I structure a rule for my behavior in some unknown future situation, then I'm saying I can't trust my ability to make a good decision based on the specific circumstances at that time and incorporating what I learned from this experience.

CHAPTER 11

Self-Responsibility

Self-actualization, an outcome goal of a positive self-concept, means actualizing or becoming yourself. Only you can recognize yourself. Internal control reflects the belief that only you are able to recognize and actualize yourself, and therefore, only you can logically be *responsible* for such actualization.

Self-responsibility means being answerable and accountable for yourself—for your actions and choices. Since only you can know whether you are being and living your life in a way that is true to yourself and that you feel satisfied with, only you can be responsible for your choices and actions.

Conceptual Belief #3:

Responsibility for Choices

All of us are responsible for our own actions and choices and for making decisions regarding who we are and how we live our lives.

❖ EXERCISE 50:   SELF-RESPONSIBILITY

Who is responsible for your happiness? To what extent do you believe you are responsible for the choices and decisions you make in your life? Justify your answers.

The underlying belief of responsibility means you acknowledge that you are accountable for your choices, actions, and feelings. We often develop rationalizations to convince ourselves that we are not responsible for ourselves.

❖ EXERCISE 51:   RATIONALIZATIONS FOR UNACCOUNTABILITY

Identify rationalizations that you use to convince yourself, and others, that you are not responsible for your

choices: "I didn't have any choice."   "My Father would disown me if I didn't go."   "I'm just shy."

actions: "She made me do it."   "It wasn't my idea."

feelings: "You made me feel bad."   "You hurt my feelings."

Try practicing the following affirmation:

>*"I am able to respond in a way that reflects being true to myself. I am able to make choices and decisions for myself and my life that I feel good about."*

THE EXPECTED

The day—it dawned on schedule
And I knew just what to do.
I rose, then bathed with ritual
After jogging, one egg, and the news.

The morning was clear and cloudless
As I turned to my tasks for the day.
It was best, they had said, not to question
But delve in without delay.

My reward would come at noontime,
They said, as they filled my plate.
But the food I once found to nourish
Now left an empty, hungry ache.

The afternoon sky appeared threatening
As I returned to my task on cue.
I questioned the forecast of sunshine
Feeling distracted, distraught, and confused.

I had believed what they said and discounted
What I knew to be true all along
But it was safer to trust their judgment
Than to risk myself being wrong.

At dusk the storm—it had broken
As I stood staring out my abode.
I sank stunned, and clutched my illusion
That the storm kept on to erode.

The night found me staring at darkness
Long after the storm had gone.
I lay aching in grief at the knowing
That I can't stay where I thought I belonged.

The day—it dawned on schedule
And I know what I had to do...

—VONDA LONG, 1983

Interpersonal Communication Skills

Communication reflects and affects how we feel
about ourselves—our self-concepts.

I continue to discover that It doesn't matter so much what I say but that I feel good about saying it. It doesn't matter so much what I do but that I feel good about doing it. The process doesn't last—and the most important product is sometimes simply feeling good about the process.

CHAPTER 12

The Complexity
of Communication

Communication reflects how we feel about ourselves, and how we feel about ourselves affects our communication. Our self-concept, then, both affects, and is reflected in, our communication.

Communication is a complex process with multiple interactive variables. The better we can understand how these variables interact, the more effective we can be in our communicating, and our understanding of ourselves and others.

❖ **EXERCISE 52:   THE COMPLEXITY OF COMMUNICATION**

Have two people, a "speaker" and a "responder," stand facing each other, with six "interpreters" lined up behind the responder. Have the speaker make a statement to the responder, such as "You look good in that shirt." Before answering, the responder is to consult with each of the six interpreters, one at a time. Each interpreter represents one of the following aspects of the responder's perspective. Each builds his or her response on the previous interpreter's response, creating a cumulative perspective for the responder.

First Interpreter: *Objective interpretation of content.* This person represents the responder's objective interpretation of the content—the explicit verbal message: "The sender of the message thinks I look good in this shirt."

Second Interpreter: *Subjective interpretation of meaning.* This person identifies how the responder interprets the nonverbal, implicit message: "This sounds like a come-on," or "This sounds patronizing."

Third Interpreter: *Feeling about interpretation of meaning.* This person identifies how the responder feels about the interpretation of meaning, such as angry, embarrassed, hurt: "I feel embarrassed that this is a come-on," or "I feel angry that I'm being patronized."

221

Fourth Interpreter: *Feeling about the feeling.* This person identifies how the responder feels about the feeling: "I feel ashamed that I feel embarrassed," or "I feel scared because I feel angry."

Fifth Interpreter: *Defense mechanism to protect self from the feeling about the feeling.* This person identifies the responder's defense mechanism, such as withdrawal, attack, or avoidance, that will be used as protection: "Because I feel ashamed I'll avoid eye contact with you and try to avoid you."

Sixth Interpreter: *Actual behavioral response.* This person formulates the actual verbal and nonverbal response the responder will give: "I'll look down, mumble 'thanks' with embarrassment, and try to get away as quickly as possible."

(BASED ON SATIR, 1988)

After sequential, accumulative input from all six interpreters, the responder replies to the sender, reflecting the actual behavioral response culminating with, and identified by, the sixth interpreter.

Repeat the exercise using the same statement, but changing one or two aspects of the responder's input. Repeat the exercise putting six people behind the speaker, also. What does this exercise indicate regarding the complexity of communication?

Consider a recent interaction in which someone said something to you to which you had an emotional reaction. Analyze your reaction using the six aspects.

Speaker's statement to you

Objective interpretation of content

Subjective interpretation of meaning

Feeling about interpretation of meaning

Feeling about the feeling

Defense mechanism to protect self from feeling about the feeling

Actual behavioral response

What role does each of the six aspects play in affecting your actual behavioral response? Which, if any, would you like to change?

Self-Development Question: How do your underlying beliefs, and particularly your beliefs regarding rights, respect, and responsibility, affect your interpersonal communication?

Skill-Development Question: What does knowledge of the six aspects suggest to you about what might be going on with the person with whom you're talking?

When I feel hurt or angry in an interaction my inclination is still to hide those feelings from the other person. Afterward—when I'm safe on the other side of the situation—I usually look back and feel I could have, and would have liked to have, risked more. I see then that I could have been more assertive, could have shared more of my feelings. The feeling emerges as a sense of having betrayed my Self. I held back. I wasn't totally there and therefore I missed the experience. A part of me experienced it, but then I wonder what it would have been like had I allowed my whole Self to have been there.

CHAPTER 13

Components of Communication— Listening, Responding, and Expressing

Expressing refers to sending a message from the sender's perspective. Responding is sending a message back to someone who has already expressed a thought. Listening means focusing on receiving a message. Components of communication reflect and affect our self-concepts—how we feel about ourselves.

Consider how you listen, respond, and express interpersonally relative to

1. Your goal of communicating: What do you want to accomplish?
2. Your self-concept: What does your style of listening, responding, and expressing reflect about how you feel about yourself?
3. Your underlying beliefs: What message are you sending regarding rights, respect, and responsibility?

Although a single usage may not be particularly significant, *patterns* of usage begin to reflect and communicate powerfully. Consider your *patterns* of usage of listening styles, response styles, and styles of expressing.

LISTENING

Listening is probably the most central component of communication; however, it's probably the least well developed. There are five basic types of listening:

1. *Nonlistening*

   Nonlistening means you are not paying attention to what is being said. Nonlistening includes ignoring. Indications of nonlistening are inappropriate responses, interruptions, or no response.

2. *Pretend listening*

   Pretend listening *looks* like listening, (eye contact, open body posture, nodding), but you are actually nonlistening; you are thinking about something else. Pretend listening may happen when you are rehearsing what you are going to say, or when you are just focused on something else.

3. *Selective listening*

In selective listening you screen certain types of messages or information and pay attention to others. Selective listening can be a useful skill, used appropriately. Focusing on the facts rather than feelings, and screening out extraneous noise are examples of appropriate selective listening. Focusing only on issues that push your "buttons" or relate to *your* interests are also examples of selective listening. The key to useful selective listening is in identifying appropriate selectiveness.

4. *Self-focused listening*

Self-focused listening focuses on your perspective, as the listener. It involves judging, interpreting, and experiencing the information as it impacts you. Often, self-focused listening occurs simultaneously with selective listening. You select elements that impact you, then focus on getting an opportunity to make your own response rather than continuing to listen. Indications of extensive self-focused listening include responses that are evaluative or judgmental, and rehearsing rather than listening.

5. *Empathic listening*

Empathic listening is listening to the message and accurately understanding the *sender's* perspective. Empathy means understanding the experience—the thoughts, feelings, and/or behaviors—of the other person.

All five listening styles can be helpful at times, depending on your goal at the moment. Although you may use all types of listening to varying degrees, you have likely developed a pattern of using some more consistently than others.

❖ EXERCISE 53:   LISTENING STYLES

Consider your listening patterns in conversations. Identify the styles of listening you tend to use most. What's your goal? What do your listening patterns reflect regarding your self-concept—how you're feeling about yourself? What do your listening patterns reflect regarding your underlying beliefs concerning rights, respect, and responsibility?

Set up a dyad with a partner and solicit feedback on your listening styles. Consciously use each listening style. Which felt most familiar? What effect did each have on your partner? Consider your goal of communicating. How effective was each style in accomplishing your goal?

RESPONSE STYLES

We have all developed response styles that we use in our interpersonal communication. The following are eight common response styles:

1. *One-Upper*

   The one-upper responds in a way implying that your situation isn't very significant. No matter what you say, one-uppers can top it with a better story about themselves. "You think you have it bad? Wait until you hear *my* situation." One-uppers reflect their own perspectives. One-uppers generally reflect *control* rather than an acknowledgment of individual *rights*.

2. *Discounters*

   The discounter discredits your feelings, thoughts, behaviors, or experiences. You are not taken seriously. This can take the form of put-downs ("You don't have it as bad as you think you do"), sarcasm or joking ("Life is pretty tough all over these days"), or reassurance ("I'm sure you'll feel better tomorrow," or "You're just in a phase; you'll grow out of it"). Discounters also reflect their own perspectives. Discounters reflect *judgment* rather than *respect*.

3. *Expert*

   The expert implies a hierarchy. It may be real—parent-child, employer-employee—or imposed—based on the assumption that the expert "knows" more than you do. The experts say directly or indirectly that they have the answer and that your job is to do what they say. For example, "I want you to go home, sit down, and make a list of everything that's bothering you." Experts assume authority, can be patronizing, and give commands. Experts reflect their own perspectives. Experts can communicate *control*, *judgment*, and *rescuing*.

4. *Advice Giver*

   The advice giver needs no hierarchy or authority to tell you what you should do. The key words for an advice giver are "should" and "ought": "What you should do is…" Advice giving is one of the more common response styles. Advice givers set themselves up to assume responsibility for others (rescuing) and then to be blamed when their advice doesn't work. Advice givers can reflect *control*, *judgment*, and *rescuing*.

5. *Cross-Examiner*

   Cross-examiners respond with a series of questions that can make you feel as though you're being interviewed or "crossed-examined." The series of questions appear to be designed to get information. The underlying message with cross-examiners is that if they can get enough information, they can do something—presumably provide the "answer." Cross-examiners implicitly take control of the interaction: "So did you say anything?" "Did you get upset?" Usually, they use closed questions. Cross-examiners can demonstrate *control*, *judgment*, *rescuing*, and/or *blaming*.

6. *"Canned" Counselor*

   The canned counselor reflects an insincere focus on the other person's perspective. The words may focus on the other's perspective—"So how do you feel about that?"—but the nonverbal message suggests insincerity. "Canned" counselors can reflect *judgment* rather than *respect*.

7. *Problem Solver*

The problem solver responds in a way that says "I know what your problem is or will find out." The implied message is "you don't, or can't." The trouble with this response style is that clients never get a chance to figure things out for themselves. The problem solver can incorporate cross-examining and advice giving—"So have you tried talking directly to your father yet?" The focus is on the counselor's perspective. Problem solvers can reflect *control*, *judgment*, and *rescuing*.

8. *Empathizer*

The empathizer focuses on trying to understand the other person's perspective and on helping the client to discover and understand his or her own perspective. Empathic statements are used, such as "It sounds like you're really feeling upset about your performance and are wanting to do something to rectify it." When questions are used, they're open questions, like "What options do you see?" Empathizers focus on the client's perspective, and generally reflect *rights*, *respect*, and appropriate *responsibility*.

❖ **EXERCISE 54:   RESPONSE STYLES**

Consider your response patterns in conversations. Identify the response styles you tend to use most. What's your goal? What do your response patterns reflect regarding your self-concept and how you're feeling about yourself? What do your response patterns reflect regarding your underlying beliefs concerning rights, respect, and responsibility?

An important question to consider when assessing your typical response patterns is whether your response is helping you accomplish your goal. Consider your goal in communicating as you continue the exercise.

Form a dyad with a partner, responding as naturally as possible. Solicit feedback on your response styles and patterns. Which one or two response styles did you use most? What was the impact on your partner? Consider your goal in communicating. How effective was each response in accomplishing your goal? What message did you communicate regarding your beliefs about rights, respect, and responsibility?

STYLES OF EXPRESSING

Expressing refers to the initiation of a message that comes from the sender's perspective. It involves identifying one's thoughts, feelings, or behaviors, and communicating them. Just as we have developed patterns of listening and responding, we have also developed patterns of expressing. Four typical styles that reflect and affect our self-concepts—how we feel about ourselves—are listed below. We tend to use all these styles, but to varying degrees.

1. *Passive*

   A passive style of expressing is defined as one that does not express your perspective, thoughts, feelings, or behaviors. When you are passive, you ignore your own rights and repress your thoughts, feelings, and behaviors. A passive style reflects a lack of sensitivity to yourself. A passive style is often associated with fear.

   parameters: Passivity is dishonest, self-denying, disrespectful, and irresponsible to yourself.

   promotes: Anxiety, fear, disappointment, anger, and resentment

   motivation: Avoidance of perceived discomfort, fear of rejection, belief of unworthiness, undeservingness, lack of skills

   consequence: You don't get what you need and don't take care of yourself; you get angry, develop a poor self-concept, and feel a desire to blame others.

2. *Aggressive*

   An aggressive style of expressing is defined as one that expresses your perspective in a dominating, unnecessarily forceful, or threatening manner. An aggressive style reflects a lack of sensitivity to the person to whom you are speaking. Aggressiveness reflects a lack of acknowledgment of the other person's rights, as well as a lack of respect. Aggressiveness is often associated with anger.

   parameters: Aggressiveness is defensive, dominating, insensitive, and hostile.

   promotes: Anger, guilt, self-righteousness

   motivation: Anger, fear of not getting what you want, insecurity

   consequence: You create isolation and alienation.

3. *Indirect*

   An indirect style is defined as an expression of your perspective, thoughts, feelings, and/or behaviors through an indirect or unclear means. A sarcastic response, nonverbal expression of irritation, rolling of the eyes, heavy sigh, or taking your anger out on an innocent bystander are all examples of indirect expression. Passive-aggressiveness, placating, and manipulation are types of indirect expression.

An indirect style of expressing reflects a lack of acknowledgment of rights, a lack of respect for both self and others, and a lack of responsibility-taking. Indirect expressing is often associated with both fear and anger, as well as sadness.

parameters: Indirect expressing is dishonest, disrespectful, irresponsible, and difficult to interpret.

promotes: Fear, anger, anxiety

motivation: Fear, avoidance of confrontation, belief in your unworthiness

consequences: You may not get what you need; you may blame others and develop a poor self-concept.

4. *Direct*

Direct expressing is defined as a clear, accurate expression of your perspective, thoughts, feelings, and behaviors. Direct expressing acknowledges rights, shows respect, and reflects appropriate responsibility-taking. Direct expressing is associated with feelings of gladness.

parameters: Direct expressing is open, honest, direct, and self-enhancing.

promotes: Responsibility, rights, confidence, self-respect, positive self-concept

motivation: To accomplish goals, actualize capabilities and potential, be responsible

consequences: You get a positive self-concept and honest relationships.

❖ **EXERCISE 55: STYLES OF EXPRESSING**

Consider your styles of expressing. Identify the style you tend to use most. What's your goal? What do your styles of expressing reflect regarding your self-concept—how you're feeling about yourself? What do they reflect regarding your underlying beliefs concerning rights, respect, and responsibility?

Set up a dyad in which you express a concern, need, or request to your part-ner. After five or ten minutes, solicit feedback from your partner regarding your style of expressing. Which style(s) do you use most? What was your goal in communicating? To what extent does your pattern help facilitate that goal?

When I feel I have to think through everything so that I know what I'm going to say before I say it, it's usually because I don't trust my ability to think (or feel) as I'm dealing with the situation. Ironically—when I do that I then am less able to deal with the situation because I'm hanging on to the sequence of statements I created in the past, and am preoccupied with trying to make them fit in the present.

CHAPTER 14

Mechanisms of Communication: Verbal and Nonverbal Behavior

The components of listening, responding, and expressing refer to *what* you do in interpersonal communication; the mechanisms refer to *how* you do it. Mechanisms of communication reflect and affect self-concept—how we feel about ourselves—and can either encourage or discourage effective communication.

Consider how you use verbal and nonverbal communication interpersonally, relative to

1. Your goal of communicating
2. Your self-concept
3. Your underlying beliefs

Consider patterns of usage and the message they communicate.

VERBAL COMMUNICATION

Verbal communication comprises statements and questions. Statements are messages sent to another person; they do not directly request a response. They are *expressive* (reflecting the speaker's perspective) or *empathic* (reflecting an understanding of the other person's perspective).

Questions are messages sent to another person that directly request a response. There are two types of questions: open questions and closed questions. *Open* questions are questions that cannot be answered with "yes" or "no," such as, "How are you feeling?" *Closed* questions are questions that can be answered with "yes" and "no," such as "Are you tired?"

❖ EXERCISE 56:   VERBAL COMMUNICATION

Form a triad. Have one person share some information with you that is of concern to him or her. Respond naturally. Have the third person observe and take notes regarding your use of verbal communication: statements (expressive and empathic) and questions (open and closed). Solicit feedback, then do a self-analysis, considering the following:

237

What are your patterns of use? What's your goal? What effect do your patterns have on your goal? What do your patterns reflect regarding your self-concept and underlying beliefs? What changes might you make in your verbal communication to increase your effectiveness?

NONVERBAL COMMUNICATION

Nonverbal communication includes silence and nonverbal behaviors. Silence can be considered *effective* when it is helpful, and when one or both people are using it productively. Silence is *ineffective* when it is used to control or make a power play, or when neither person is finding it useful.

Nonverbal behaviors include all behaviors, other than verbal, that transmit messages from one person to another. These include facial expressions, body position, head nodding, gestures and body movements, and eye contact.

❖ EXERCISE 57:   NONVERBAL COMMUNICATION

Repeat the triad exercise above. Do a self-analysis of your nonverbal communication. What patterns of nonverbal behavior do you have? What's your goal? What effect do your patterns have on your goal? What do your patterns reflect regarding your self-concept and underlying beliefs? What changes might you want to make to increase your effectiveness?

CONSONANCE

Consonance refers to consistency of verbal and nonverbal messages. A statement, "I'm really happy," that is accompanied by a smile reflects consonance; the same statement said through clenched teeth reflects a lack of consonance. Consonance reflects self-concept and how you're feeling about yourself.

❖ EXERCISE 58:   CONSONANCE

Using the interaction from the above exercise, do a self-analysis of your consonance.

PART

V

A Stage Framework
for Personal Growth

Change is inevitable; growth is optional.

—ANONYMOUS

I'm beginning to understand that sharing and revealing myself Doesn't mean sharing any particular thing—But rather means just being who I am—At the time. As I go.

CHAPTER 15

A Stage Framework

Regardless of whether you are attempting to facilitate growth toward the development of a positive self-concept for yourself or others, the process can be conceptualized as involving the same three basic stages. These stages include the following:

STAGE 1.   THE PRESENTING PROBLEM

In Stage 1, the focus is on the presenting problem—what you perceive initially as the concern or issue. The presenting problem is the answer to the question "Why are you upset?" Typically, the presenting problem focuses on an external event—e.g., partner leaving, failing an exam, losing a job. The following steps can assist in clarifying the presenting problem:

1. **Telling the Story:**   *Describe the situation, issue, or concern. What's your experience?*

    E.g., I discovered my husband was having an affair and I confronted him. He denied it. I raged at him.

2. **Initial Awareness:**   *Describe your thoughts, feelings, and behaviors. How do you feel, what do you think about, and how are you reacting to what's going on?*

    E.g., I'm angry; I think my husband's a jerk and I'm raging at him.

3. **Initial Problem Identification:**   *Identify the problem. Why are you upset, frustrated, concerned, or unhappy?*

    E.g., My husband left.

STAGE 2.   THE UNDERLYING ISSUES

In Stage 2, the focus shifts from what was initially identified as the presenting problem, which is typically an external focus or event, to the underlying internal issues. Stage 2 explores the presenting problem to discover why it is of concern or problematic to you. Typically, the underlying issues relate to how you experience and feel about yourself—feeling unlovable, feeling inadequate, feeling like a failure. The following steps can assist in clarifying the underlying issues:

1. **Problem Differentiation:** *Describe what is your problem and what isn't. Identify the external events or circumstances that are disappointments for you, and identify the internal issues that you want to do something about.*

    E.g., What my husband does is my husband's problem. The effect it has on me is my problem. My husband's having an affair is a disappointment. My feeling rejected and unlovable is an internal issue.

2. **Examining the Issues:** *Describe what makes this issue a problem for you. How does it relate to the components of a positive self-concept?*

    E.g., When I feel rejected and unlovable, I defend myself by getting angry and raging. I do not love and accept myself.

3. **New Perspectives and Insight:** *Describe how this response relates to your happiness, self-fulfillment, or satisfaction with your life. What's the real problem?*

    E.g., I feel rejected and unlovable because I don't love and accept myself. I feel unworthy of love.

STAGE 3. DIRECTION AND CHANGE

Having gained new perspective and insight in Stage 2, your focus in Stage 3 shifts to direction and change. Direction and change address your choice of action or inaction as a result of your new self-understanding—becoming more accepting, going back to school, becoming more assertive. The following steps can assist in facilitating direction and change:

1. **Direction:** *Describe what you want to do as a result of your new perspective.*

    E.g., I want to begin nurturing myself—doing for myself what I've been wanting others to do for me.

2. **Implementation:** *Describe how you are going to do it.*

    E.g., I'm going to make a list of things I find nurturing, like going to a good movie, taking candlelight bubble baths with classical music, having a massage, and start doing one a day for myself. I'm going to schedule this on my calendar.

3. **Change:** *Do it.*

The following chapters provide self-development exercises that correspond to the steps in each of these three stages. Exercises are intended to promote personal growth toward the development of a positive self-concept. They can be used in promoting your own growth or in facilitating the growth of others.

Whenever I find myself thinking of being able to tell others what I did so that they will know I had a good time, I know that I am probably not having one.

CHAPTER 16

Stage 1:
The Presenting Problem

The presenting problem is what you initially perceive to be the concern, issue, or problem. It can be clarified through telling the story, initial awareness, and initial problem identification.

❖ EXERCISE 59: TELLING THE STORY

Choose a situation that you are frustrated or upset about. Describe the situation issue or concern. What's your experience of it?

❖ EXERCISE 60:   INITIAL AWARENESS

Describe your thoughts, feelings (e.g., mad, glad, sad, scared), and behaviors. How do you feel about, what do you think about, and how are you reacting to what's going on?

❖ EXERCISE 61:   INITIAL PROBLEM IDENTIFICATION

Identify the problem. Why are you upset, frustrated, concerned, or unhappy?

Describe the extent to which the problem you've identified is external, some-
thing outside yourself—an event, another person's action, a result, or an
outcome.

My experience has shown me that in forcing myself to do what does not feel right for me at the time, I do not enjoy the experience nor do I do well at the activity.

CHAPTER 17

Stage 2: Underlying Issues

Stage 2 involves exploring the underlying issues. Why is the initial problem identified in Stage 1 an issue for you? How does it relate to your self-concept?

Because our goal is the development of a positive self-concept—or feeling satisfied with who we are and how we're living our lives—a "problem" will be defined as something that interferes with feeling satisfied with who we are and how we're living our lives.

Because feeling satisfied with who you are is based on an *internal* judgment and perspective, only another internal judgment or perspective can become an interfering problem. For example, if I judge myself to be a failure because I failed an exam, it is my *internal perspective* that is the problem, not my failing the exam.

We are conditioned to view external events as the problem (failing the exam); therefore, exploring underlying issues involves a process of distinguishing between *external events* and *internal issues*.

External events include everything outside ourselves: the anger of a friend, the deadline of a boss, the grades of a son, the filing for divorce by a partner. We have no direct control over external events; we may have influence, but not control. External events have no direct relationship to how we feel about ourselves. It is how we internally view and judge the event and, subsequently, ourselves, that become the problem.

Internal issues include everything within ourselves: our thoughts, feelings, and behavior—that is, how I feel about myself, how I judge myself, how I choose to act. We have direct control over internal issues. Internal issues relate directly to how we feel about ourselves and how we're living our lives.

Problem differentiation is the process of distinguishing between external events and internal issues. External events will be considered "disappointments" and internal issues will be considered the real "problem" or underlying issue. In terms of personal growth, accurate and honest problem differentiation is probably half the solution.

❖ **EXERCISE 62: PROBLEM DIFFERENTIATION**

Use the Problem Differentiation Model in Figure 17.1 to analyze a recent event about which you are/were upset. Begin with the external event and translate it into an internal issue. Determine (1) what elements are unreasonable "shoulds" and expectations that you might want to let go, and (2) what changes you might want to make to feel better about yourself. This procedure is merely a guide to help differentiate between external disappoint-

ments and internal issues. You are the judge. Use the example in Figure 17.2 as a guide.

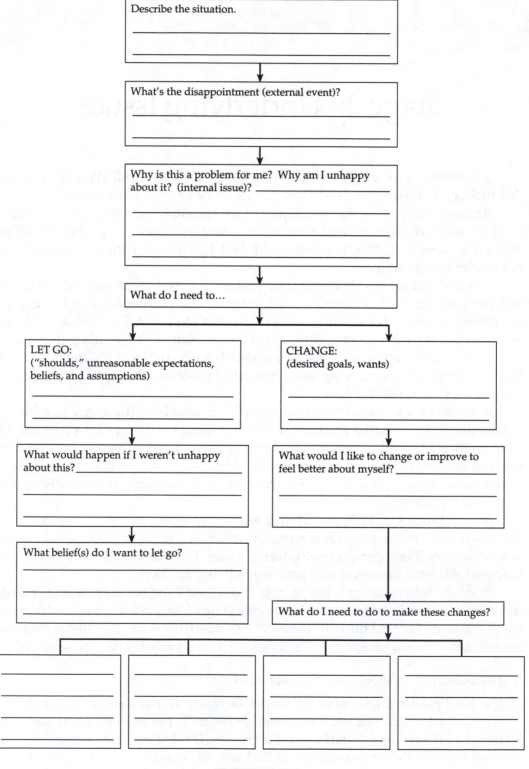

FIGURE 17.1
Problem Differentiation Model

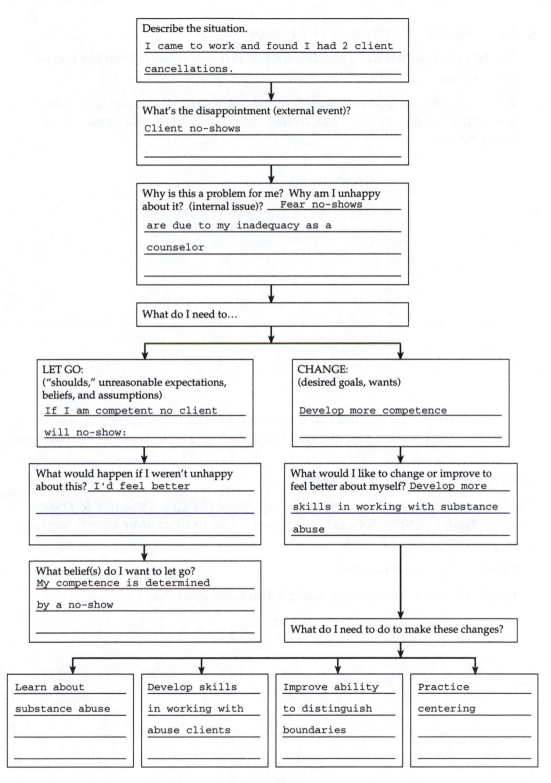

FIGURE 17.2
Problem Differentiation Model: Example

❖ **EXERCISE 63:   EXAMINING THE ISSUES**

Identify the external disappointments and internal issue(s) from the previous exercise.

E.g., external disappointments: client no-show, client not using opportunity to grow; internal issues: feeling inadequate, fear of incompetence

The goal of examining the issues is to gain a better understanding of what the internal issues are so that you can gain insight into the changes you want to make.

❖ **EXERCISE 64:   NEW PERSPECTIVES**

Identify new perspectives you've had regarding changes you'd like to make in yourself. Consider both (1) beliefs you'd like to discard or change, and (2) changes you'd like to make.

Beliefs to discard or change:

E.g., If I have a no-show, it means I'm incompetent.

Changes in behaviors:

E.g., I'm going to learn more about substance abuse and working with substance-abuse clients.

The past is simply the present that was and the future simply the present that will be. Sometimes it seems so simple: The easiest way to feel good about the past and future is to feel good about the present.

Stage 3:   Direction and Change

Direction and change involve deciding what you want to *do*, or not do. Once you have gained new perspectives and insight in Stage 2, the next step involves deciding what changes you want to make, and changing. This includes determining direction (What do you want to do?), developing an implementation strategy (How are you going to do it?), and making the actual behavioral change (Doing it).

Determining direction involves making a decision about which way to go. In developing a positive self-concept, direction determination typically involves letting go an external focus (e.g., self-worth based on external recognition and attention) and taking responsibility for the internal issue (e.g., wanting to feel important and cared about). In other words, direction determination involves making a decision to stop doing behavior that takes you in a direction you don't want to go, and start doing behavior that takes you in a direction you do want to take.

One way to help determine direction for behavioral change is to consider behaviors that interfere with an identified goal to be self-defeating behaviors. The following exercise will guide you through a self-assessment of self-defeating behaviors that may be interfering with the direction you want to take.

❖ **EXERCISE 65:   DIRECTION DETERMINATION: SELF-DEFEATING BEHAVIOR (SDB) ANALYSIS**

1. *Choose one goal that has emerged as a result of the exercises in Stage 2.*

   E.g., speaking assertively about my feelings or opinion when I want to

2. *Identify a self-defeating behavior that interferes with that goal. If it doesn't interfere with your goal, it's not self-defeating.*

   E.g., freezing, not saying anything

3. *Identify the fear that keeps you from reaching your goal. There are three basic fears:*

      a.   Fear of what might happen to me—E.g., rejection, getting hit or hurt, being criticized.

      b.   Fear of what I might find out about myself—if I try I might find out I have no talent or aptitude.

      c.   Fear of the unknown.

E.g., If I say how I feel and what I think, you may not like me.

4. *When I do this self-defeating behavior (SDB), what do I think would happen to me if I didn't do it? (What's the fear?)*

E.g., By not freezing up, but speaking out, I might fumble my words, embarrass myself, and be ridiculed and rejected.

5. *When I do this SDB, I'm sacrificing a part of myself. What am I sacrificing (think about your goal)?*

E.g., In not speaking up, I'm sacrificing my right to be myself, my belief in my ability to speak up, and my responsibility to and for myself. I'm sacrificing being able to learn, through practice, how to better speak up for myself.

6. *Some behavioral techniques I use that help perpetuate this SDB include these (what do you do that allows you to continue this SDB?):*

E.g., avoiding situations when I think I might have to speak up. When a situation arises when I might have to speak up, I leave. Sometimes I feign a coughing spell so I won't have to speak.

7. *If I chose to work toward my goal instead of practicing the SDB, the behaviors I would do would be:*

E.g., seeking out situations when I might have to speak up. Staying and speaking for myself instead of leaving or coughing.

8. *Sometimes I try to convince myself, with rationalizations, that I do not have control over my SDB. Tactics I use include:*

E.g., rationalizing that "I can handle it better than they can, so I won't say anything to upset them." "This is the natural way of dealing with it." "I learned it from my family." "It's hereditary." "It's just the way I am." "I'm a Taurus, and that's how we are."

9. *Consider each of your rationalizations. Would you rather keep the rationalization or would you rather have your goal?*

10. *When I do my self-defeating behaviors I may pay prices. I pay prices in terms of how I feel about myself. I pay prices physically, emotionally, intellectually, and socially. In the circles in Figure 18.1, identify the prices in each area, which, in turn, cause you to pay more prices. Use the example in Figure 18.2 as a guide.*

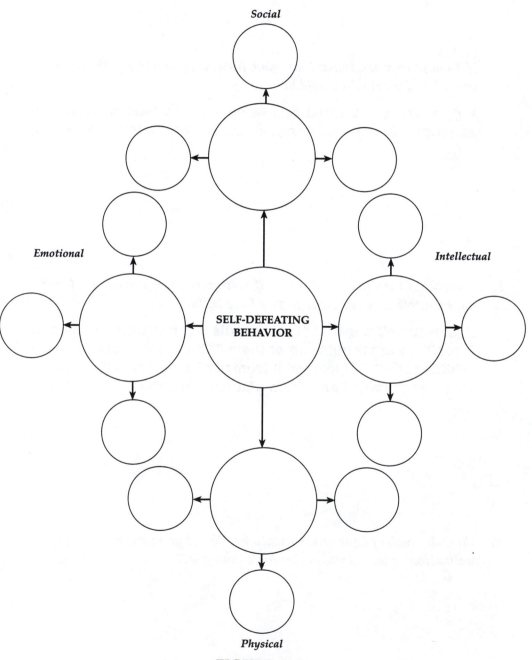

FIGURE 18.1
Prices Exacted by Self-Defeating Behaviors

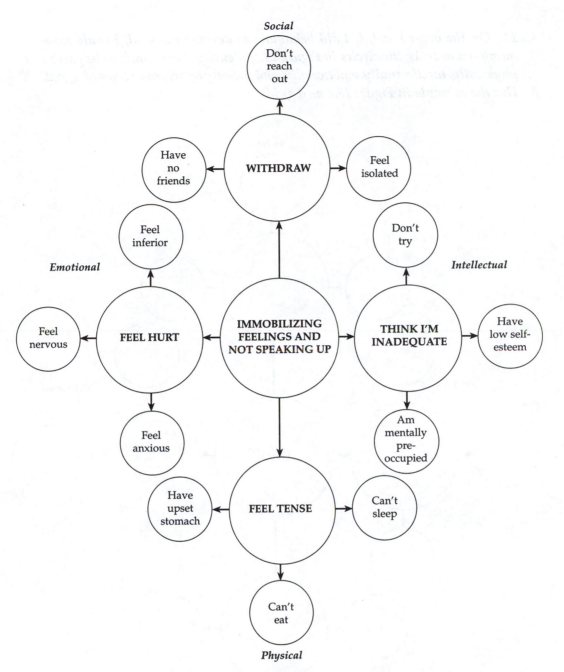

FIGURE 18.2
Prices of Self-Defeating Behaviors: Example

11. *On the other hand, if I did behaviors to develop my goal, I could have many rewards. In the circles in Figure 18.3, identify the rewards to be gained physically, intellectually, emotionally, and socially by working toward a goal. Use the example in Figure 18.4 as a guide.*

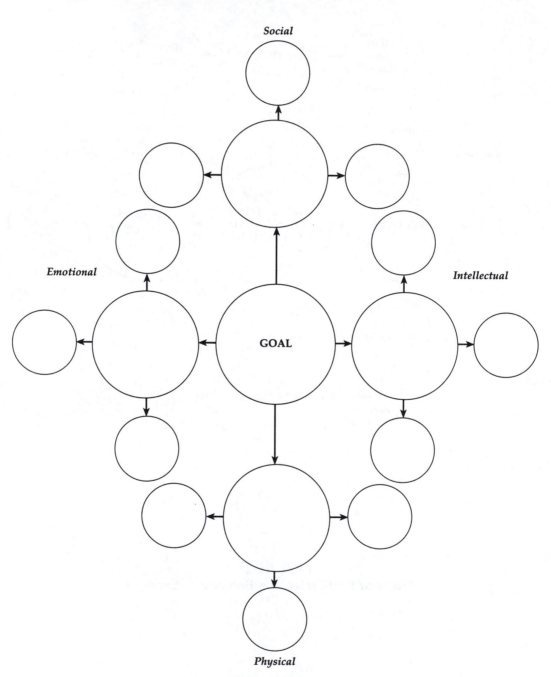

FIGURE 18.3
Rewards of Goal-Oriented Behaviors

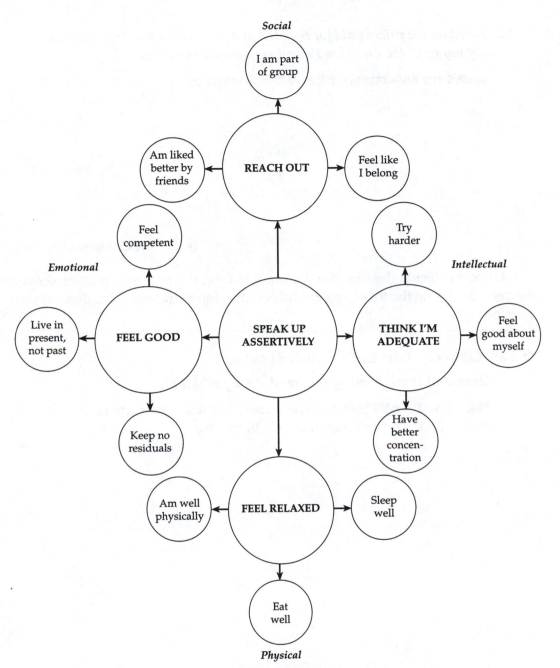

FIGURE 18.4
Goal-Oriented Behaviors: Example

12. *Based on the prices paid for the SDB and potential rewards for develop-ing my goal, the direction I'm going to decide to take is:*

speak up assertively when I feel I want to.

(based on Chamberlain, 1981)

Once we decide the direction we want to take, the next step is to implement the decision. Using the following guidelines, develop a strategy to implement your decision:

❖ Exercise 66:   Implementation Strategy

1. *Use* small *steps. Identify the* small *steps you'll take.*

   E.g., I will (a) identify how I feel, (b) acknowledge to myself how I feel, and (c) say out loud how I feel.

2.   *Be* Specific *regarding what you're going to do. Identify here* specifically *what you plan to do.*

I will say first to myself, then I will write, then I will say out loud how I am feeling.

3.   *Be* Positive. *Focus on what you're* going to do, *not what you're* not going to do. *What are you going to do?*

I'm going to say to myself how I feel, as opposed to simply saying I'm not going to run.

4. *Be* Repetitive. *The small, specific steps should be ones that can be repeated easily over and over. Having something simple enough to repeat in many different situations is important. Assess your plan:*

    E.g., "I'm going to acknowledge to myself how I feel" can be repeated in many different circumstances.

5. *Be* Independent. *The strategy must be one that you can carry out alone. You must not be dependent on someone else to be able to do it. Assess the independence of your plan:*

    E.g., I'll say to myself how I'm feeling. I won't rely on my friend to tell me whenever she thinks I'm not saying how I'm feeling.

6. *Be* Reasonable. *Unless the plan makes sense to you, it won't work. Assess the reasonability of your plan.*

    E.g., simply acknowledging how I feel to myself and saying it aloud sounds reasonable to me; trying to get any particular reaction doesn't sound reasonable.

After determining the direction you want to go (goal) and developing a strategy for implementation, the final step is to actually *do* the behavior: make the change. To do this involves a commitment to change. If you have trouble implementing the change, (1) reassess whether it is really a change you want, or (2) try organizing a more disciplined approach. This can involve scheduling time, or enlisting the help of a support network (friends or an organized support group).

❖ **EXERCISE 67:   CHANGE**

Reassess whether this change (goal) is something you really want. If so, try approaching it with the same commitment you have to going to work. For example:

1. Schedule a time each day when you are committed to doing it
2. Prioritize it (perhaps you can do it *first* every day)
3. Enlist support in doing it
4. Enlist feedback
5. Set aside time each day to evaluate your progress

Every so often I find myself going through a day or week doing what I have scheduled on my calendar with a focus on getting these things done. And then I remember that in focusing just on getting them done, I seldom get to enjoy doing them.

CHAPTER 19

Integration and Evaluation

Our goal has been facilitating personal growth toward a positive self-concept. A positive self-concept means you feel satisfied with who you are and how you're living your life. Remember that you alone are the judge of how you feel about yourself and how you're living your life. You alone know whether you are happy with yourself and your life. You alone can determine changes you need to make to improve how you feel about yourself and your life. You alone can change.

❖ EXERCISE 68:   A POSITIVE SELF-CONCEPT REVIEW

Summarize (1) your present self-concept, (2) underlying issues affecting your self-concept, and (3) changes you'd like to make. Consider your perspective on your being, *including beliefs regarding your* right *to be yourself,* congruence *and* self-acceptance; doing, *including your self-*respect, *competence, and* self-esteem; *and* choosing, *including beliefs regarding self-*responsibility, internal control, *and* self-actualization.

I don't want to look back on my life and wish I had given myself permission to have lived it in a way with which I was happy. If now is not the time to begin being who I want to be and living in a way that reflects personal truth and integrity, then when will it be?

REFERENCES

American Counseling Association. (1988). *Ethical standards* (rev. ed.). Alexandria, VA: Author.

American Psychological Association. (1992). *Ethical principles of psychologists and code of conduct.* Washington, DC: Author.

Arkoff, A. (1988). The meaning of personal growth. In A. Arkoff (Ed.), *Psychology and personal growth* (3rd ed., pp. 331–338). Boston: Allyn & Bacon.

Association of American Colleges. (1986). *The campus climate revisited: Chilly for women faculty, administrators, and graduate students.* Washington, DC: Author.

Bandura, A. (1969). *Principles of behavior modification.* New York: Holt, Rinehart & Winston.

Bem, S. (1974). The measurement of psychological androgyny. *Clinical Psychology, 42,* 155–162.

Brammer, L. M. (1988). *The helping relationship: Process and skill* (4th ed.). Englewood Cliffs, NJ: Prentice-Hall.

Broverman, I., Broverman, D., Clarkson, F., Rosenkrantz, P., & Vogel, S. (1970). Sex-role stereotypes and clinical judgements of mental health. *Journal of Consulting and Clinical Psychology, 34,* 1–7.

Chafetz, J. (1988). *Feminist sociology.* Itasca, IL: F. E. Peacock.

Chamberlain, J. (1981). *Eliminate your SDB's.* Provo, UT: Brigham Young University.

Corey, G., (1991). *Theory and practice of counseling and psychotherapy* (4th ed.). Pacific Grove, CA: Brooks/Cole.

Covey, G. S. (1990). *The seven habits of highly effective people.* New York: Simon & Schuster.

Cushman, P. (1990). Why the self is empty. *American Psychologist, 45* (5), 599–611.

Farrell, W. (1986). *Why men are the way they are.* New York: McGraw-Hill.

Gayton, W. F., Sawyer, B. L., Baird, J. G., & Ozman, K. L. (1982). Further validation of a new measure of machismo. *Psychological Reports, 51,* 820–822.

Hancock, E. (1981). Women's development in adult life. *Dissertation Abstracts International, 42* (6).

Heuscher, J. E. (1992). Mythology—the self-Peer Gynt. *American Journal of Psychoanalysis, 52,* 79–92.

Long, V. O. (1983a). On life and love. *American Mental Health Counselor's Association Journal, 5,* 192.

Long, V. O. (1983b). Seasons of freedom. *The School Counselor, 30,* 405.

Long, V. O. (1986). Relationship of masculinity to self-esteem and self-acceptance in female professionals, college students, clients, and victims of domestic violence. *Journal of Consulting and Clinical Psychology, 54,* 323–327.

Long, V. O. (1989). Relation of masculinity to self-esteem and self-acceptance in male professionals, college students, and clients. *Journal of Counseling Psychology, 36* (1), 84–87.

Long, V. O. (1996). *Communication skills in helping relationships: A framework for facilitating personal growth.* Pacific Grove, CA: Brooks/Cole.

Løvelie, A. (1982). *The self of the psychotherapist: Movement and stagnation in psychotherapy*. Oslo-Bergen-Tromso: Universitetsforlaget.

Maslow, A. (1968). *Toward a psychology of being*. New York, NY: D. Van Nostrand.

May, R. (1989). *The art of counseling*, (rev. ed.). New York: Gardner Press.

Mischel, W. (1966). A social learning view of sex differences in behavior. In E. E. Maccoby (Ed.), *The development of sex differences* (pp. 56–81). Stanford, CA: Stanford University Press.

Pearson, J. C. (1985). *Gender and communication*. Dubuque, IA: Wm. C. Brown.

Perls, F. (1969). *Gestalt therapy verbatim*. Menlo Park, CA: Real People Press.

Pleck, J. (1983). *The myth of masculinity*. Cambridge, MA: MIT Press.

Rogers, C. R. (1980). *A way of being*. Boston, MA: Houghton Mifflin.

Sadker, M., & Sadker, D. (1981). *The cost of sex bias in schools: The report card*. New York: Longman.

Satir, V. (1988). *The new peoplemaking*. Mountain View, CA: Science and Behavior Books.

Schaef, A. W. (1985). *Women's reality*. San Francisco, CA: Harper & Row.

Schau, C., & Heyward, V. (1987). Salary equity: Similarities and differences in outcomes from two common prediction models. *American Educational Research Journal, 24,* 271–286.

Smith, R. H., Diener, E., & Douglas, H. (1989). Intrapersonal and social comparison determinants of happiness: A range-frequency analysis. *Journal of Personality and Social Psychology, 56,* 317–325.

Spence, J. T., Helmreich, R., & Stapp, J. (1975). Ratings of self and peers on sex role attributes and their relation to self-esteem and conceptions of masculinity and femininity. *Journal of Personality and Social Psychology, 32,* 29–39.

Tannen, D. (1990). *You just don't understand*. New York: Ballantine Books.

Webster's New Collegiate Dictionary. (1981). (8th ed.). Springfield, MA: G & C Merriam.

Williamson, M. (1992). *A return to love*. New York: Harper Collins.

Woody, H. R., Hansen, C. J., & Rossberg, H. R. (1989). *Counseling psychology: Strategies and service*. Pacific Grove, CA: Brooks/Cole.

TO THE OWNER OF THIS BOOK:

We hope that you have found *Facilitating Personal Growth in Self and Others* useful. So that this book can be improved in a future edition, would you take the time to complete this sheet and return it? Thank you.

School and address: _____

Department: _____

Instructor's name: _____

1. What I like most about this book is: _____

_____

_____

2. What I like least about this book is: _____

_____

_____

3. My general reaction to this book is: _____

_____

4. The name of the course in which I used this book is: _____

_____

5. Were all of the chapters of the book assigned for you to read? _____

    If not, which ones weren't? _____

  6. In the space below, or on a separate sheet of paper, please write specific suggestions for improving this book and anything else you'd care to share about your experience in using the book.

_____

_____

_____

_____

_____

_____

Optional:

Your name: _____ Date: _____

May Brooks/Cole quote you, either in promotion for *Facilitating Personal Growth in Self and Others*, or in future publishing ventures?

Yes: _____ No: _____

Sincerely,

Vonda Long

Brooks/Cole is dedicated to publishing quality publications for education in the human services fields. If you are interested in learning more about our publications, please fill in your name and address and request our latest catalogue, using ths prepaid mailer.

Name: _____

Street Address: _____

City, State, and Zip: _____

FOLD HERE

FOLD HERE